# I AM THE CHANGE

# CHANGE

*Barack Obama and the Future of Liberalism*

## CHARLES R. KESLER

**BROADSIDE BOOKS**
*An Imprint of* HarperCollins*Publishers*
www.broadsidebooks.net

To the Memory of William F. Buckley, Jr.

HarperCollins books may be purchased for educational, business, or sales promotional use. For information, please write: Special Markets Department, HarperCollins Publishers, 10 East 53rd Street, New York, NY 10022.

Broadside Books™ and the Broadside logo are trademarks of HarperCollins Publishers.

A hardcover edition of this book was published in 2012 by Broadside Books, an imprint of HarperCollins Publishers.

FIRST BROADSIDE BOOKS PAPERBACK PUBLISHED 2013

Designed by Renato Stanisic

Library of Congress Cataloging-in-Publication Data has been applied for.

ISBN: 978-0-06-207302-0

13 14 15 16   OV/RRD   10 9 8 7 6 5 4 3 2 1

PRAISE FOR

# I AM THE
# CHANGE

"There's plenty to read on Barack Obama, and plenty to read on liberalism. In this imaginative and perceptive book, Charles Kesler brings them together, deepening our understanding of both. Kesler illuminates the contemporary crisis of liberalism—and points a way forward, up from liberalism."

—William Kristol, editor, *Weekly Standard*

"What does Barack Obama's reelection as president mean for the future of the United States and for the president's place in history? Both liberals celebrating his electoral victory, and conservatives who fear for the future of the country, may be surprised by the thoughtful, serious, and provocative answer to this question in *I Am the Change: Barack Obama and the Future of Liberalism*."

—Charles Kadlec, Forbes.com

"Brilliant yet sober, this is the deepest analysis we have had."

—Harvey C. Mansfield,
professor of government, Harvard University;
senior fellow, Hoover Institution

"Charles Kesler is the reader for whom Obama has long been asking—in the sense of 'asking for it'—and this book is the examination of the One we've been waiting for."

—Ramesh Ponnuru, *National Review*

"In his important new book, Charles Kesler provides a conservative reading of the Obama presidency that is neither dismissive nor conspiratorial. Instead, he sees Obama as the latest presidential exemplar of a liberal political tradition that for a century has worked to transform the American constitutional order in ways that would make it unrecognizable to the Founding Fathers. . . . Clearly written, patiently excavating the words and actions of Wilson, Roosevelt, Johnson, and Obama, the book is a model of intellectual history that is at the same time completely of the moment."

—Brian Anderson, editor, *City Journal*

"Charles Kesler is one of the nation's preeminent scholars on the Founding. He has now turned his attention to analyzing the liberal, left-wing, and progressive departure from it. *I Am the Change* is important reading for anyone interested in the future well-being of the United States of America. If you're going to read one book about Obama, this should be the one."

—William J. Bennett,
author of *America: The Last Best Hope*,
and former U.S. secretary of education

"Charles Kesler has produced a timely and exceptionally well-written book, full of insight, erudition, and wit. It's a must-read . . . for conservatives of all stripes—and for liberals interested in honestly examining the intellectual underpinnings of their political faith."

—Jeffrey H. Anderson, *Weekly Standard*

"Seldom have I encountered such a solid and perhaps even incontrovertible thesis, done with such great skill and learning, so soberly, and yet . . . entertainingly. Sort of like listening to Laurence Olivier doing *Hamlet*. This is not a bear-baiting book, but a brilliant analysis of Barack Obama's thought that is at once an education in political philosophy and an elegant illumination of the moment." —Mark Helprin, author of *A Soldier of the Great War*

# Contents

# Preface to the
# Paperback Edition

When I finished writing *I Am the Change*, the election of 2012 was nine months away. I made no predictions about the contest, though I made no secret of my preference for Mr. Obama's conservative opponent, whoever he or she would turn out to be. Whether Obama won or lost, I foresaw problems for liberalism. But a sufficiently convincing reelection victory would, I argued, breathe new life into the liberal movement, create fresh opportunities for its immoderation, and tempt the president into yet loftier expressions of self-regard.

So, the time is ripe for a little political stock-taking.

By 2012, hope and change had lost their savor. It was Obama's opponents who now pleaded for change, and the incumbent, no phrasemaker despite his reputation as a speech-maker, reached back into the Left's shopworn slogan bag for a fusty rallying cry: "Forward." The only surprise was the punctuation mark at the end. To younger voters, the dot signified an internet life-form, a promise of sites and suffixes to come—forward.org, perhaps, or more likely, www.whitehouse.gov/forward. This was not your father's Forward, in short, or your grandfather's, or any of those other countless exhortations to rush to the barricades that earned the word its Wikipedia definition as a "generic name of socialist publications."

To everyone else, the period was supposed to signal the administration's resolute matter-of-factness: it said, simply, there is no alternative to our kind of progress. You can't turn back the clock, so it's Forward, period; or as the Clintons, back when they were the new kids in town, might have put it, it's Forward, stupid. And yet it was no time at all before the campaign had stealthily replaced the attention-getting period with an exclamation point. At the Democratic National Convention in Charlotte, exclamation points were everywhere! The return of the old leftist battle cry "Forward!" seemed to signal either a new candor or a nagging doubt about victory. Maybe the voters thought there was a choice in this election, after all. The times were not particularly favorable to the president. Economic recovery had been slow, though increasingly steady; unemployment had dropped from 10 percent to around 8 percent, and housing prices were rebounding. In his acceptance speech, Obama acknowledged his administration's "mistakes" but appealed to "We the People" for absolution and a renewed faith in his vision for America. Like Lincoln, he noted, he was mindful of his failings and limitations. (When it comes to humility, he's tops.) The election "four years ago wasn't about me," he contended implausibly. "It was about you. My fellow citizens—you were the change." But "if you buy into the cynicism that the change we fought for isn't possible," he warned, "well, change will not happen. . . . Only you have the power to move us forward." Or to put it less delicately: If I lose it will be your fault, you ingrates.

On election day the people came to a fork in the road and, in

Yogi Berra's immortal words, took it. Despite the mood of most postelection commentary, the results were neither a landslide for President Obama nor a wipeout for the GOP. Judging by the numbers alone, the public decided to march not forward but sideways. To be sure, the president's 332 electoral votes amounted to a comfortable victory. Measured against all the presidential contests since 1896, however, the year which political scientists often regard as the beginning of "modern" American politics, Obama's winning percentage (61.71 percent of the electoral vote) was below average, ranking twenty-second out of thirty. His percentage of the popular vote (51.06 percent) put him eighteenth among the thirty victors. More remarkably, he lost ground compared with his election in 2008, when he captured 52.87 percent of the popular vote. The political scientist who calculated these rankings, Jack Pitney of Claremont McKenna College, points out that in races with a serious third-party contender the winner's percentage of the popular vote will be unusually lowered (as happened to Bill Clinton in 1992 and 1996). To compensate, one ought to look in addition at the winner's margin of victory in the popular vote. By that metric, Obama's winning *margin* over Mitt Romney (3.85 percent) was quite ordinary and indeed far below average—only twenty-fourth out of thirty, about half of his 7.3 percent victory over John McCain four years before.

In short, this was no electoral earthquake. The Republican Party lived to fight another day; and it retained control of the House of Representatives. Divided government, the status quo ante, was the electorate's decree again. At the state level, Republicans added a governorship and kept control of a majority of state legislatures. The same polls that in the final week or so showed (accurately, it turned out) Obama pulling safely ahead showed Romney with a notable lead two or three weeks prior. It was a more fluid, and more even, race than the final results suggested.

Yet in politics as in life, numbers don't tell the whole story. The election was still six weeks away when *Newsweek* (now defunct as a print publication, though I'm sure there's no connection) proclaimed on its cover that Obama was "The Democrats' Reagan." In the cover essay, Andrew Sullivan speculated that Obama "will emerge as an iconic figure who struggled through a recession and a terrorized world, strafing the ranks of al-Qaeda, presiding over a civil-rights revolution, and then enjoying the fruits of the recovery." Sullivan was on to something, insofar as Obama is indebted for his rhetorical sensibility to two modern presidents above all, Bill Clinton and Ronald Reagan. From Clinton he learned to praise capitalism and pass the collection plate, to distance himself from the excesses of Sixties radicalism, and to obfuscate liberalism's new programmatic rights as forms of "opportunity." From Reagan he learned to speak in optimistic and patriotic tones about the country, without (please note) conceding anything in principle to Reagan's own principles, or to the Gipper's conservative agenda. As a self-styled pragmatist, Obama couldn't help noticing that Clinton's rhetoric had itself aped Reagan's, and that Reagan's rhetorical leadership had over several decades earned the supreme pragmatic compliment: it worked. Obama's cheeky praise of Reagan early in the 2008 campaign had conceded that, and only that, point: "I think Ronald Reagan changed the trajectory of America in a way that Richard Nixon did not and in a way that Bill Clinton did not. He put us on a fundamentally different path because the country was ready for it."

Other liberal commentators leaped on the Reagan-Obama bandwagon, too. What Sullivan saw, which they didn't, was that Obama had always been playing a long game, aiming at fundamentally transforming the country. In fact, Sullivan underestimated Obama's zeal for transformation, which, as this book argues, owes more to the ideological afflatus of his liberal predecessors, Lyndon B. Johnson, Franklin D. Roosevelt, and Woodrow Wilson, than to any inspiration from Reagan. *Time* magazine (not defunct . . . so far) had come closer to the truth when after his 2008 triumph it put

Obama on its cover as the new Franklin Roosevelt, complete with pince-nez and cigarette holder, accompanied by the headline "The New New Deal." Obama's Second Inaugural Address was, in this respect, such a textbook example of progressivism that it ought to be reprinted as an appendix to *I Am the Change*. The speech is thoroughly Rooseveltian, steeped in FDR's 1932 and 1936 campaign appeals, and intended to establish the Democrats as the party of true Americanism and to abominate the Republicans as anti-American oligarchs and bigots. Tactically, the speech was a riposte to the Tea Party, which in its haphazard way had tried to associate the American Revolution with *opposition* to Obama. On the contrary, Obama insisted, *he* stood for the tradition of Jefferson and Lincoln. It was his opponents who were the (economic) royalists.

Strategically, then, the Second Inaugural confirmed Obama's effort, chronicled herein, to reverse the Reagan Revolution, to cast the conservative movement into the outer, undemocratic darkness, and to restore the Democrats as the majority party and liberalism as America's official faith. To begin to justify his own transformations, he had to re-define the whole American political tradition, especially the Founding, just as FDR had done memorably in the early New Deal. Thus Obama began by saluting "the enduring strength of our Constitution" (not its wisdom or justice) and affirming "the promise of our democracy," meaning the country as it might be, the America of our imagination, which to a modern liberal is the only thoroughly justifiable object of patriotic sentiments. If one believes with, say, the Reverend Jeremiah Wright that America began in principle as a criminal conspiracy by rich white men against Indians, blacks, women, the poor, and everyone else, then of course one cannot look backward or even upward for moral guidance. The only way open is Forward.

Obama then quoted the electric passage from the Declaration of Independence, "We hold these truths to be self-evident, that all men are created equal. . . ." A sentence later —one sentence!—and the Declaration was in the rearview mirror and we were off on "a never-ending journey" to "bridge the meaning of those words with the realities of our time." Only urgent, imperative actions such as approving gay marriage, fighting climate change, and protecting entitlements, he announced, "will lend meaning to the creed our fathers once declared." Like American government, the American creed is broken and desperately in need of a bailout, which liberalism is only too happy to supply. By borrowing the latest liberal priorities, and genuflecting before the latest liberal pieties, we could make the old Declaration and Constitution relevant to the present age.

Would that be the Age of Obama, as his admirers whispered after the reelection had sunk in? The hardback edition of this book bore the subtitle *Barack Obama and the Crisis of Liberalism*. With Obama's victory, is the crisis over? I venture he thinks so. He had diagnosed a kind of near-term crisis of liberalism in his 2008 campaign and in *The Audacity of Hope*, his precampaign book from 2006. As Obama saw it, the crisis was that liberals had lost confidence in liberalism. Reagan had so dominated the 1980s that his legacy, Reaganism, set the terms of public debate and public policy for two more decades. Bill Clinton's attempted challenge to the conservative status quo—his signature policy that somehow sported his wife's signature, Hillarycare—never got out of committee in a Congress dominated by his own party. After that, it was all school uniforms, midnight basketball, triangulation, welfare reform, and most galling of all to Obama, Clinton's declaration that "the era of big government is over." Liberals feared that liberalism had been neutered, or at least tamed. Obama did not. He thought that something like a new New Deal or a new Great Society—breathtaking political change along a wide front—was still possible, if only liberals would rediscover their faith in the future, if progressives would believe once more in progress. That's what Hope and Change were all about. Obama in 2008 persuaded liberals to hope once more for big political and social change; but it would all have been for naught if he hadn't delivered, which, unlike Bill Clinton, he did. In the first two years of his presidency, his forces overran the conservative lines and pushed leftward in a series of dramatic, party-line victories on the stimulus bill, Dodd-Frank's reregulation of the financial sector, and above all health care. Crisis over, he likely concluded.

Except that his legislative gains were never matched by the kind of broad and deep partisan realignment triggered by the New Deal. As a result, the GOP's counterattack in 2010 succeeded on almost every front. He regained some ground in 2012, but no one would compare Obama's political position today with the preeminence enjoyed by Roosevelt and the Democrats from 1932 to 1938. The best that John Judis, Ruy Teixeira, and other supporters can do is to predict an invincible Democratic majority that will emerge . . . sometime in the future, based on extrapolations from the growth rate and voting behavior of key Democratic constituencies—young people, single women, single mothers, gays, Hispanics, and so forth. They may be right, but it's far from certain. And by continually looking over the next hill for this new majority, they admit backhandedly that even all of Obama's considerable political skills, a deep and prolonged economic downturn, and the roiling political battles of the past decade have not sufficed to bring it into existence.

This persistent failure may not lessen Obama's self-congratulation, but it should. In his second term, count on him to try to nudge the national conversation, and eventually the national agenda, farther to the left, partly out of conviction and partly out of the need to find new wedge issues to spur the political realignment he seeks. The coy "evolution" of his position on gay marriage will be matched on other issues like taxes, a federal right to child care, and economic redistribution. Though his administration has won many legislative victories, none can be said, at least so far, to be *politically* decisive, and so the maneuvering goes on. Obama has revived liberalism, but its long-term health is far from assured. Even as defined by him, the crisis of liberalism is not over.

When the publisher suggested that the paperback edition carry a new, postelection subtitle, *Barack Obama and the Future of Liberalism*, I assented readily because whatever else the future of liberalism may have in store, it will include the crisis of liberalism, though in a different and larger sense than the president envisions. Soon modern American liberalism will come to a crossroads. Today's welfare state and, a fortiori, tomorrow's, cannot be sustained on the present tax base. Taxes on everyone, especially on the middle class, which is where the money is, will have to rise dramatically, and additional forms of taxation will have to be imposed. Nearly half, or perhaps more than half, of the national economy will have to be conscripted by government at all levels in order to keep the welfare state afloat, effectively socializing the economy. Or . . . the welfare state will have to be pared back and rededicated to helping the truly needy in ways consistent with natural rights, limited government, and the consent of the governed. It's far from clear that liberalism has the moral and political resources to reform, much less to ax, programs that it has defended for generations now as unbreakable civic promises, essential to securing ever evolving kinds of human dignity.

Complicating the dilemma will be the growing suspicion, already visible among the professoriate and other avant-garde thinkers, and glimpsed even in President Obama's glib dismissal of "absolute truth," that liberalism itself is nothing but liberals' will-to-power dressed up in its Sunday-go-to-meeting best. That sentiment, if shared by their political leaders, would exacerbate the dangers along either path liberals might take. And the American people can hardly be expected to keep believing in the justice of liberalism if liberals can't or won't. On both fiscal and philosophical grounds, then, I think the crisis of liberalism, far from being over, is only beginning.

# Introduction

**B**arack Obama had the distinction of being the most liberal member of the United States Senate when he ran for president in 2008. The title had been conferred by *National Journal*, an inside-the-beltway watchdog that annually assigns senators (and congressmen) an ideological rank based on their votes on economic, social, and foreign policy issues. Obama was more liberal than the Senate's independent socialist, Bernie Sanders, considerably more liberal than Barbara Boxer or Harry Reid, and dramatically to the left of his opponent in the primaries, Hillary Clinton. It was only one ranking, but it captured something important. He was much more liberal than his presidential campaign let on.

Since then, we have learned a lot more about his political leanings as a young man, which were fashionably leftist, broadly in keeping with the climate of opinion on the campuses where he found himself—Occidental College, Columbia University, Harvard Law School. As a senior at Columbia, he attended the 1983 Socialist Scholars Conference, sponsored by the Democratic

Socialists of America. It met in Manhattan at the Cooper Union, site of one of Abraham Lincoln's major speeches, but this conference commemorated not the Great Emancipator but Karl Marx, on the centenary of his death. Though a meeting of democratic socialists and, yes, community organizers, the conference as well as his long-running friendships with radicals of various sorts would have drawn more sustained attention if the Cold War were still raging. But it was not, and Obama pleaded youthful indiscretion and drift; and of course his campaign did its best to keep the details from coming out. He still had to answer, in some measure, for his ties to William Ayers and Jeremiah Wright, but the issue with, say, the good Reverend concerned his sermons about race and Middle East politics, not his penchant for visiting and honoring Fidel Castro, not to mention the Marxist Sandinistas in Nicaragua.[1] Partly by avoiding the worst of the old anti-Communist gantlet, Obama became the most left-wing liberal to be elected to national executive office since Henry Wallace.

Still, the President is not a self-proclaimed socialist—nor like Wallace, a self-deceived fellow traveler or worse. Obama never went so far, so openly—whether out of inertia, political calculation, or good sense—and therefore never had to make a public apostasy. As a result, we know less about his evolving views than we might like, though probably more than he would like. He calls himself a progressive or liberal, and we should take him at his word, at least until we encounter a fatal contradiction. That's only reasonable and fair; and it avoids the desperate shortcut, gratifying as it may be, of unmasking him as—take your pick—a third-world daddy's boy, Alinskyist agitator, deep-cover Muslim, or undocumented alien. Conservatives, of all people, should know to beware instant gratification, especially when it comes wrapped in a conspiracy theory. In any case, hypocrisy, as Rochefoucauld wrote, is the tribute that vice pays to virtue, and Obama seems to think it would be a virtuous thing to have been a lifelong liberal, even if he wasn't. And so the question arises: what does it mean anymore to be a liberal?

This book is about Barack Obama and his place in modern American liberalism. It approaches liberalism as he sees it, as a form of progressivism, which is more or less how its greatest twentieth-century champions understood it, too. (Capital-P Progressivism hereafter refers to the movement that arose in the first two decades of the past century; small-p progressivism to the more general belief in an inevitably glorious, man-made future. Progressivism was progressivist; but not every form of progressivism, e.g., Marxism, was compatible with Progressivism.) Neither a biography of Obama nor a history of liberalism, this volume focuses on liberalism's essence—what it is, where it came from, where it's going—and how the president sees himself in that picture. Foreign policy, as such, figures very little here, not because liberals don't have interesting and highly untraditional views on the subject but because their views of domestic policy, and particularly of the grounds and purposes of political life, are more fundamental. For similar reasons, we don't consider the broader American Left, spanning labor unions, social reform movements, the Socialist and Communist parties, and the like, but confine ourselves to the main political and intellectual developments and the most prominent and ambitious political leaders—the great men of the age, as Woodrow Wilson might put it.[2] Most liberals would recognize Wilson, Franklin D. Roosevelt, and Lyndon B. Johnson as the most eminent, certainly the most consequential of elected liberal statesmen, even though they are lamentably dead, white, and male. Teddy Roosevelt has admirers, including, lately, Obama; but though he managed to be many things, often at the same time, TR was never a Democrat nor a sentimental egalitarian.[3] Doubtless John F. Kennedy is more beloved than LBJ, but he doesn't hold a candle to the uncouth Texan in terms of building the modern welfare state and fulfilling the civil rights agenda. Truman's achievements are mostly in foreign policy, and many of them, like launching American participation in the Cold War, reek too much of the Vietnam War for contemporary liberal tastes. Bill Clinton, well, has his own problems. Ditto Jimmy Carter, though not the same ones.

When liberals tell their own story, they emphasize the unplanned, improvisational character of what came to be called liberalism. As Eric Alterman, a professor of English and journalism who likes to write books defending liberalism, declares, "liberalism arose as a matter of pure pragmatism with next to no theory in the first place and was led by a politician [FDR] who prided himself on his willingness to try almost anything. . . ." This argument, repeated in countless mainstream histories, presupposes that socioeconomic change is the driving force, and that politics, at least good, liberal politics, is a kind of reaction—an adjustment of governing institutions and policies to the changing realities of society. Liberalism comes across as defensive and modest, in fact downright conservative, but also inevitable. Political change can't lag behind social change for long, and what liberals do is simply mind the gap: they prescribe the minimal adjustments necessary to keep the social organism healthy and whole. The story has the advantage of de-radicalizing liberalism, and of distracting attention from its actual ideas and from their role in its real genesis and growth. It's the equivalent of the policeman saying, Move along, folks, there's nothing to see here. Keep moving. . . . The same liberals who push this pragmatic account invariably speak at the same time of their movement's "ideals" or "vision," revealing that liberalism as conservative adjustment cannot be the whole truth. In fact, both arguments, for liberalism as slow change and for liberalism as the hope of idealistic or radical transformation, were originally made by the same man, Woodrow Wilson, when he helped to found modern American liberalism. Franklin Roosevelt was a young man then, a Wilsonian progressive serving in his leader's administration.

Wilson thought the modest story about liberalism partly true, but partly a noble lie to cover the remarkably thorough break he intended to make from the original, and still more or less prevailing, interpretation of the principles of American politics. The cover-up was thus coeval with the crime, you might say. Without dismissing the liberal gift for moderation and capacity for compromise, this book will shine a light on the peculiar radicalism inherent in American liberalism ever since its origins in the Progressive movement. Liberalism was a choice, not a destiny, and in its rise to power ideas, not material conditions, were in the forefront. Its "foundational" ideas, as we say today (alas), when seen in broad daylight, point up the connections between the several installments of liberal reform, as envisioned and explained by the leading liberal reformers themselves. Pay no attention to the man behind the curtain, commands the Wizard of Oz. In this case, however, the famous liberal statesmen are well in front of the curtain, exposed as only presidents can be, and their ideas and their reasoning are open to anyone who knows how to read. Our method, accordingly, is to let these renowned liberals speak for themselves as much as possible. We stick as closely as we can to their own words from their speeches, books, and letters. We try to understand them as they understood themselves, before criticizing or evaluating them—though this book, written by a conservative, doesn't hesitate to criticize, taking as its touchstone the very precepts the liberals were gently but firmly trying to supplant, the principles of Abraham Lincoln and the American Founders.

Good-natured liberals may be surprised by this frank account of their own creed. Contrary to what they've probably read, liberalism was all about theory, and a kind of theory much more hostile to American premises than they've been told. But what is this "pure pragmatism" they are supposed to be celebrating instead? Pragmatism was a new development in American philosophy, a late-nineteenth-century school that modestly disclaimed ultimate truths, abstract theories, and final ends in favor of a method that seeks and finds truth in "what works." At the time, this meant what works according to the methods of the natural sciences, particularly Darwinian biology, and the new social sciences modeled on the natural. To the Pragmatists themselves, like John Dewey and William James, their approach was revolutionary, or to be precise, a revolution against the old ways of philosophy and politics that had discovered supposedly permanent truths like natural rights, the laws of nature and of nature's God, and unchanging species. In other words, Pragmatism was itself a theory; and to be a Pragmatist was already to incline against some of the main ideas of American constitutionalism. This isn't the blancmange that the term suggests to most people today. As for philosophers, few would adhere nowadays to old-style Pragmatism anyway, having long since exchanged it (see Chapter Five) for a more radical formulation. "What works" now implies not Dewey's "method of intelligence," but rather what works for you—for your favorite values, self-created lifestyle, or will to power. Have liberals noticed that they are in danger of becoming confidence men, selling belief in what Obama sometimes calls "universal truths" that they know are not universal and suspect strongly are not true?

The sense of liberalism as something novel, audacious, and comprehensive has faded, lost in the enduring authority of its innovations, the familiarity of its claims, and its temporary taming by the Reagan Revolution. Whatever else he has accomplished, Obama has reminded conservatives and liberals alike that liberalism can be an aggressive doctrine unashamedly pursuing the transformation of the country. Its robustness shouldn't surprise. Its transformations are relatively recent. The twentieth century was, as the late Thomas B. Silver used to say, "the liberal century." Conservatism was a late arrival, debuting as a self-conscious intellectual movement only in the 1950s, and lacking significant political success until the 1980s. By contrast, the liberal storm was already gathering in the 1880s, and broke upon the land in the new century's second decade. It had made deep, decisive changes in American politics long before conservatism as we know it came on the scene. Those who would like to limit or reverse liberalism's damage must face the fact that already, over several generations, it has pervasively reshaped Americans' expectations of government and of life. Nonetheless, it didn't win these victories all at once. Modern liberalism spread across the country in three powerful waves, interrupted by wars and by rather haphazard reactions to its excesses. This fact is encouraging, because it shows that it can be stopped, and discouraging, because it hints it cannot be stopped for long. But then conservatives and moderates—even liberals—haven't had a commanding view of the movement in a long time, and so the past is not necessarily prologue.

Each wave of liberalism featured a different aspect of it—call them, for short, political liberalism, economic liberalism, and cultural liberalism—and each deposited on our shores a distinctive type of politics—the politics of progress, the politics of entitlements, and the politics of meaning. These terms are conceptual rather than, strictly speaking, historical. They help to organize our thinking more so than our record-keeping, inasmuch as elements of all three were mixed up in each stage. Although it wasn't inevitable that one wave should follow the next, a certain logic connected the New Freedom, the New Deal, and the Great Society. Each attempted to transform America, as their names suggest, and the second and third waves worked out themes implicit in the first, which is why the book devotes so much attention to Progressivism in Chapter Two. But the special flavor of each period owed much to the issues and forces involved, the legacy of previous reform, the character of the political leaders, and the disagreements within and between the generations of reformers. The third wave, centered on the Sixties, showed just how fratricidal liberalism could become.

The first and most disorienting wave was political liberalism, which began as a critique of the Constitution and the morality underlying it. That morality, Wilson charged, the natural rights doctrine of Thomas Jefferson and Lincoln, was based on an outmoded account of human nature, an atomistic and egoistic view that needed to be corrected by a more well-rounded or social view, made plausible by the recent discovery that human nature was necessarily progressive or perfectible. So-called natural rights were actually historical or prescriptive, evolving with the times toward a final and rational truth. The eighteenth-century Constitution, based on the eighteenth-century notion of a fixed human nature with static rights, had in turn to be transcended by a modern or living constitution based on the evolutionary view. Drawing on a curious and unstable mixture of Social Darwinism, German idealism, and English historicism, Wilson outlined the new State that liberals would ever after be building, the goal of which would be nothing less than man's complete spiritual fulfillment. Though he insisted that the process of change was seamless, he was candid enough, especially in his political science but also in his popular speeches, to explain the metamorphosis he intended.

The second wave explicitly adopted the name of liberalism, laying aside the old banner of Progressivism. It championed liberality or generosity in the form of a new doctrine of socioeconomic rights, and tried to connect the new rights to the old, the Second Bill of Rights (as FDR called it) to the First. Instead of rights springing from the individual, the New Deal reconceived individualism as springing from a new kind of rights, created by the State. The new entitlement-style rights posed as personal rights, even though they effectually attached to groups; but due to the slight family resemblance, they allowed Roosevelt to present himself and the New Deal as the loyal servants and successors of the American Revolution, of the old social compact suitably updated. Liberalism's third wave, cultural or lifestyle liberalism, hit in the 1960s. It was only when this wave crashed around them that the radical character of liberalism became clear to the American people; only then that conservatism became, at least temporarily, a majority movement, insofar as it stood for America against its cultured despisers and reformers. The Great Society agreed with the New Deal that government had to provide for Americans' necessities in order that they may live in freedom, but it denied that freedom from want and freedom from fear (along with freedom of speech and worship) were any longer sufficient for all-around human liberation. Freedom required not merely living comfortably but also creatively, a demand that the New Left took several steps further than poor Lyndon Johnson was willing or able to go.

In the Sixties the "peculiar" character of the radicalism bound up with contemporary liberalism began to tear it apart. Modern liberalism had always been a complex blend of an evolutionary rights doctrine, pursuing greater equality of conditions among individuals and groups so that they would have greater freedom to realize themselves or choose their own lifestyles; a unified State directed by experts in public administration and social and economic development who would enforce the conditions of social morality; a philosophy of history that assured that rational progress was possible and indeed inevitable; and a faith in political leaders who could envision history's next move and inspire the people to march into that better future. The second and third elements depended on hierarchy, education, and authority of a surprisingly conservative sort. The first and fourth were more fluid, open-ended, and relativist. When social morality collided with personal liberation, and the State's authority clashed with the people's rights, and the assumptions of rational progress were denied by protestors who preferred to make history by following their authentic selves rather than admire history as it came to an end—then liberalism began to unravel. For conflicting reasons, liberals lost faith that they were on the right side of history, and that the State could ever provide the conditions for complete self-development or spiritual fulfillment.

Obama inherited that frayed liberalism. Against long odds, he's tried to reunite its dissonant parts and restore its political élan. He brought America to the verge of a fourth wave of political and social transformation, something that neither Democrats nor Republicans thought possible. But as the latest embodiment of the visionary prophet-statesmen he hasn't been able to sustain the deep connection to the American people that his election in 2008 seemed to promise, and that his desire to restore liberalism as the country's dominant public philosophy required. Perhaps after the debacle of the Great Society, three decades in the political shadow of Ronald Reagan, and the current protracted economic doldrums, Americans have grown suspicious of the liberal vision of the future as a kind of Brigadoon—a land of wonders that voters glimpse every four years but that fades quickly into the mists, and from which no one has ever returned. Unlike any of his liberal predecessors, Obama's tortuous doubts about American exceptionalism lead to a sense of his estrangement from his own country, a disability not relieved by his profession, in Berlin, that he is a citizen of the world as well. By American standards he seems to lack both the citizen's pride and the immigrant's gratitude.

Nonetheless, the health care reform bill alone could plausibly establish him as the savior and renewer of the liberal political faith for its second century. But the program's life is threatened on all sides. The Supreme Court could deprive him of this triumph by roughly striking down Obamacare; and if it doesn't the people themselves may do so, indirectly, in the 2012 election and its aftermath. Nor can one forget that rich countries around the globe are being bankrupted by their government's debt and deficits, just as he vows to expand ours by leaps and bounds. Obama's fate is tied to liberalism's, and vice versa. It was precisely a hundred years ago that Woodrow Wilson launched modern liberalism in all its hubris. That year, 1912, was also the year of the *Titanic*. Does anyone see an iceberg ahead?

**1**
\_\_\_\_

# The Audacity of Barack Obama

**M**artin Luther King Jr. had already been killed in Memphis in April, and Bobby Kennedy nine weeks later in Los Angeles, when the Democratic National Convention pulled into Chicago in August 1968, intent on nominating Hubert Humphrey for president. By then Americans were inured to seeing their cities burn: almost a hundred riots had broken out after King's murder, a striking case of shooting the messenger and then his message—of nonviolence. But none of those explosions or any of the others since the 1965 Watts riots had the shock value of the melee in Chicago between the peace activists and the police. This wasn't a protest against racial inequality or against a brazen assassination, nor even rioting mainly for fun and profit.[1] The demonstration and the ensuing confrontation in Grant Park were emphatically political, a revolt against the Democratic Party and its old guard (with its old rules: Humphrey was the last Democrat to get his party's nod without entering a single presidential primary), a battle between the center-left and the farther left for the liberal soul. *Götterdämmerung*, an astute participant later called it: the twilight of the gods. Inside the International Amphitheatre, Senator Abraham Ribicoff, in the course of nominating the antiwar successor to the recently martyred Kennedy, declared that "with George McGovern as president of the United States we wouldn't have to have Gestapo tactics in the streets of Chicago." Not to be outdone, Mayor Richard J. Daley shook his fist and loosed a perfect storm of politically incorrect epithets in the Connecticut senator's direction, denouncing Ribicoff in terms that might curl even current Chicago mayor Rahm Emanuel's toes.[2] For five days the battles raged inside and outside the convention. Haynes Johnson, who covered it as a reporter, summed up the results four decades later: "In its psychic impact, and its long-term political consequences, it eclipsed any other such convention in American history, destroying faith in politicians, in the political system, in the country and in its institutions."[3]

To be sure, Haynes Johnson was not an unbiased observer, and so one must adjust his terms. By "politicians" he meant *liberal* politicians, and by "political system" he meant the liberal political system as it had developed since the New Deal. Chicago did not destroy American conservatives' faith in *conservative* politicians, in conservative ideals, or in the enduring institutions of the country. If anything, the events in Chicago confirmed the right wing's opinion of American liberalism's advanced state of decomposition. What the 1968 Democratic National Convention bloodily exposed was the civil war within liberalism itself. At the center of that conflict was the Vietnam War, which the protesters and New Left groups condemned not merely as unjust but sick. Increasingly, they regarded Vietnam as the outward and visible sign of an inward and spiritual disgrace in the American soul: this was a country that *enjoyed* waging cruel and unjust wars because this was a cruel and unjust country, whose Gestapo tactics at home perfectly reflected its fascist warmongering abroad. Their enemy was not merely the war but "the war machine." In this sense Johnson got it exactly right. What was at stake was nothing less than the moral and political respectability of American civilization, and especially of America regarded, as Mayor Daley and millions of his fellow Democrats regarded it, as the bulwark and very home of liberal decency. And since liberalism was in the saddle in those days and the Democratic Party predominant, Johnson can be forgiven for thinking everyone perceived the same catastrophe he did.[4]

Still, Americans of every political persuasion would have been struck by the contrast with Grant Park forty years later on Election Day, 2008. Thousands of citizens of every race and background streamed happily down Michigan Avenue for hours before the victory rally began, before they could be sure (though hardly anyone doubted) there would be a victory to celebrate, converging around the dignified stage where Barack Obama would acknowledge his election as the forty-fourth president of the United States. To millions in America and around the world, it felt historic, but in a good way. *Time* magazine's post-election cover story began: "Some princes are born in palaces. Some are born in mangers. But a few are born in the imagination, out of scraps of history and hope." (The birthers turned the bliss into a joke: Why doesn't Obama have a birth certificate? Because he was born in a manger!) *Time*'s writer, the talented Nancy Gibbs, didn't specify which kind of royalty had been elected president but implied . . . a Prince of History and Hope. She noted, for example, that Obama disclaimed personal ambition. "*I'm not the one making history,* he said every chance he got. *You are.*" Nonetheless, "people were waiting for him, waiting for someone to finish what a King began." She meant Martin Luther King Jr. It's remarkable how much the day's elation drew on the hope, or the hype, that the curse of the Sixties had finally been extirpated, King's dreams for the country realized. Camelot, civil rights, youthful idealism, an end to unjust wars, the certitudes of American progress, the banishment of tragedy—all came together again, this time in perfect harmony. "*Remember this day,* parents told their children as they took them out of school to go see an

African-American candidate make history," observed Gibbs. "An election in one of the world's oldest democracies looked like the kind they hold in brand-new ones, when citizens finally come out and dance, a purple-thumb day, a velvet revolution. . . . You heard the same phrases everywhere. *First time ever. In my lifetime. Whatever it takes.*"[5]

David Remnick, editor of the *New Yorker* and soon-to-be Obama biographer, agreed that Chicago was an epochal moment in American race relations. "The color line had not been erased or even transcended, but a historical bridge had been crossed," he wrote. *The Bridge*, his life of Obama, elaborated on the metaphor, spanning the story of American race relations from "Bloody Sunday," March 7, 1965, in Selma, Alabama, to Obama's victory speech in Grant Park. Remnick relayed a story from Roger Wilkins, who asked his white neighbor Ann why she was supporting Obama. "She looked at me and said, 'Because I want to feel good about my country.'" Ann's feeling was widely and passionately shared. When the Electoral College numbers signaled victory, the new president and his family finally appeared onstage, greeted by what Remnick called "well-mannered pandemonium: crying, flag-waving, the embracing of friends and strangers." "It's been a long time coming, but tonight, because of what we did on this day, in this election, at this defining moment" declared Obama, "change has come to America."[6]

Michael Tomasky, the editor of *Democracy: A Journal of Ideas*, whose principal idea is democratic liberalism, captured the moment this way:

> The image of Barack Hussein Obama speaking to America from his stage in Grant Park that night in November 2008 as president-elect was, for liberals, one of the most staggering images we've ever seen. One felt—many millions of us felt—almost invincible in a way; finally

justified in our beleaguered beliefs, after so many years of
despondency and rage; aware in fresh and unprecedented
ways of our collective power, like mortals transformed
into superheroes in the movies, realizing for the first time
that they could fly or crush stone. . . . All things seemed
possible.[7]

It wasn't just stardust, in other words, or even a racial breakthrough that all Americans could be proud of, that the president-
elect represented. There was something about "the image of Barack
Hussein Obama speaking to America" that filled millions of his
fellow liberals not so much with pride in their country's progress
as with a sense of "collective power," like supermen to whom suddenly "all things seemed possible." Race shrank to a subplot in
the drama of *Yes We Can*, "that timeless creed," "that American
creed," as Obama called it, exhorting the crowd with the phrase
seven times. His was no mere personal, racial, or even partisan triumph, but a long-overdue renewal of the progressive cause, which
in his view was the American cause. If not exactly two sides of the
same coin, he and American liberalism yet were intimately connected, dependent on the same "staggering images" of a political
hero with superhuman abilities who could, by the magic of his
own rhetoric, inspire belief in a future in which all things desirable seemed possible. The alternative to such "invincible" power
was, in Tomasky's terms, hideously bleak: a future, like the recent
past under George W. Bush, of "despondency and rage," of liberal
beliefs "beleaguered," unfulfilled, nugatory. Grant Park 2008 was
not so far from the rage of Grant Park '68, after all. But now the
dreams of the antiwar protesters, the McGovernites, the radicals,
and to a certain extent even the old Democrats had come true, or
at least were on the verge of being fulfilled, thanks to Obama's victory. He seemed to have reconciled the Sixties antagonisms, to have
restored the house of liberalism to its accustomed and ruling place

in American politics. On that day and for many months thereafter, through his own audacity and vision, Barack Obama defined the power and the possibilities of American liberalism. Only gradually would liberalism come to define—to circumscribe and finally endanger—the power and the possibilities of Barack Obama.

## Nemesis

As a candidate, Obama used to joke about the sun breaking through the clouds when he started to speak. For the past year and a half, at least, the sun god has deserted him. His legislative accomplishments peaked in early 2010 with the passage of his health care and financial reform bills. Both are less popular now than then, and polls consistently show a majority of Americans opposed to the Patient Protection and Affordable Care Act. The economic stimulus bill failed to fulfill the administration's promise to keep unemployment below 8 percent; on the contrary, unemployment soared to over 10 percent and lingered painfully in the 9s and high 8s before drifting lower. Obama's own approval numbers plunged from 69 percent in early 2009 to the low 40 percent range, before stabilizing in the mid-to-high 40s. What his secretary of the Treasury, Timothy Geithner, dubbed the "recovery summer" of 2010 disappeared beneath the waters of the BP oil spill in the Gulf of Mexico—a recovery operation of a different, and protracted, sort. Day after day those belching black clouds mocked Obama's billowing oratory, becoming an omen of subterranean nemesis for this president who had promised to heal the earth and calm the seas. The low point was reached when, criticized for his excessive deference to BP's team and the government's own experts in the Gulf, the president asserted he was looking for some "ass to kick." He added that of course he first had to consult the experts, including his Nobel Prize–winning energy secretary, to determine *whose* ass to kick!

It's not surprising that by the second summer of Obama's term, supporters like Tomasky had begun to write essays with titles like

"Against Despair." "It seems likely," he admitted, "that American liberals will never again for the foreseeable future feel quite like we did that night" in Grant Park. Euphoria always leaves a hangover, and this was a doozy. "American liberalism has . . . been living through a painful period of coming to terms with [the] reality" that "all things weren't possible," Tomasky explained, and that in retrospect it was a bad idea to "insist on thinking of Obama . . . as liberalism's redeemer."[8] The despondency deepened when the November 2010 election results rolled in. Amid the worst economy in a generation and in some respects since the Great Depression, the only populist movement to emerge was a right-wing one, the Tea Party. Aided by the Tea Party, the Republicans won historic gains in the House of Representatives and in state legislatures, picking up more than 60 seats in the House and almost 700 at the state level, and electing six new GOP governors (including in the key swing states of Michigan, Ohio, Pennsylvania, and Wisconsin) and six senators. Independent voters went overwhelmingly for GOP candidates. For two years the Democrats had enjoyed undivided control of the national government, holding all three of the elective branches. The midterm elections ended that. After this "shellacking," as he called it, the president formerly known as Prince would no longer be able to dictate the terms of debate, and would have only the pleasures of partisan gridlock, a sluggish economy, and golf games with Speaker John Boehner to look forward to.

## A New Majority

Tempting as it might be to write President Obama off, it would be a big mistake. Whatever else he may accomplish, his staggering victory on health care reform has earned him a future place on the Mount Rushmore of liberalism, alongside those other supreme hero-statesmen of the creed, Woodrow Wilson, Franklin D. Roosevelt, and Lyndon B. Johnson. Assuming that his signature achievement is not unceremoniously repealed and replaced, Obama will almost certainly become one of the Democratic immortals, the giants who built and expanded the modern liberal state. Unlike his predecessors, he will deserve credit for bringing liberalism back, at least temporarily, from a period of prolonged and profound decline. After Ronald Reagan's presidency, for almost twenty years liberalism seemed an exhausted political faith, ashamed of its own name—in 2008 both Hillary Clinton and Obama preferred to be called "progressives"—and surviving as a political force only by tacking rightward. From the very beginning of his campaign for the presidency, however, Obama set his sights not on modest policy gains but on bold, systemic changes to energy policy, environmental regulation, taxation, foreign policy, and of course health care, all designed, in his words, to *remake America*, and all presupposing that "shock-and-awe statism," as Indiana governor Mitch Daniels termed it, was again possible.

In pursuit of that vision, Obama led his party in 2008 to a commanding victory, winning the presidency with a majority of the popular vote—something a Democrat hadn't managed since Jimmy Carter's squeaker thirty-two years before. It's worth recalling the sweep of the Democrats' win only four years ago. In individual terms, Obama's was the biggest triumph for his party since LBJ's landslide in 1964—though that understates Obama's achievement because Johnson was already president then. Among *non-incumbent* Democrats, Obama's presidential victory ranks an astonishing third in U.S. history, behind (as a percentage of the popular vote) only Roosevelt's in 1932 and Andrew Jackson's in 1828.[9] In short, Obama already compares favorably to the greatest Democratic politicians of the nineteenth and twentieth centuries. And though he didn't sweep into office as many congressmen and senators as FDR or LBJ did in their big years, he helped boost the Democrats' margin of control in both houses of Congress, which had been liberated from a decade of GOP dominance in 2006. They moved from 235 to 257 seats in the House of Representatives, and from 51 (counting the two independents) to 59 in the Senate. Those larger margins proved vital to the passage of the stimulus, Obamacare, and the financial reform bill.[10]

Liberals now may carp at him for losing the initiative or not fighting hard enough, but Obama is playing a long, high-stakes game, and it's not at all clear he's losing. After all, he sees himself engaged in an epic struggle: he is trying to reinvigorate the very possibility of liberalism, as both moral aspiration and political program. To understand what he's aiming at, it helps to see the problem he is trying to overcome. Consider, then, his own view of the parlous state the Democrats were in before he came along.

In *The Audacity of Hope*, published in 2006, his second autobiography and first campaign book (focused nominally on his U.S. Senate years, not quite two of them at that point) and the source of his most thoughtful campaign speeches, he treats the party elders respectfully, but not exactly warmly. He mentions Teddy Kennedy three times, calling him one of the Senate's best storytellers; devotes a page to Al Gore's emotions after his "precipitous fall"; and acknowledges "the Kerry people" who invited him to speak at the 2004 Democratic convention. Obama goes out of his way to emphasize he was a newcomer to the party who couldn't even get a floor pass to the 2000 convention. Reflecting on the elections of 2000 and 2004, he confesses, "I sometimes felt as if I were watching the psychodrama of the Baby Boom generation—a tale rooted in old grudges and revenge plots hatched on a handful of college campuses long ago. . . ."[11]

He praises Bill Clinton more highly than any other contemporary Democrat, because Clinton recognized the staleness of the old political debate between Left and Right and came close to moving beyond it with his politics of the Third Way, which "tapped into the pragmatic, nonideological attitude of the majority of Americans." But Clinton blew it, and the author gradually explains why and how. First, he regrets Clinton's "clumsy and transparent" gestures to the Reagan Democrats, and his "frighteningly coldhearted" use of other people (for example, "the execution of a mentally retarded death row inmate" before a crucial primary). Then Obama notes sadly that Clinton's policies—"recognizably progressive if modest in their goals"—had commanded broad public support, but the president had never been able, "despite a booming economy," to turn that support into a governing coalition. Finally, he gently charges Clinton with the worst offense of all: strengthening the forces of conservatism. Due to Clinton's untidy life story, which Obama doesn't hesitate to summarize—"the draft letter saga, the marijuana puffing, the Ivy League intellectualism, the professional wife who didn't bake cookies, and most of all the sex," not to mention the "undeniable evidence" of such "personal lapses"—Clinton prepared the way for George W. Bush's victory in 2000.[12]

In his campaign and presidential speeches, Obama can't afford to be so candid—he still needs Hillary and Bill's supporters, after all—but he subtly makes his point. For example, in his acceptance speech in Denver, the single biggest speech of the 2008 campaign, he laid at Bill Clinton's feet the oldest backhanded compliment in the books, thanking the former president "who last night made the case for change as only he can make it. . . ." That's a subtle double insult. It reminds the discerning ear of Clinton's characteristic bloviation, and then of his larger political failings—that is, when you think back on Clinton, you're reminded why the Democrats so desperately need Obama.[13]

Granted, Obama holds Clinton to higher standards than he does the other party elders. Jimmy Carter, Gore, Kerry—these mediocrities lacked the political talent that Clinton squandered, in Obama's estimation, and they were innocent of political daring. Their shortcomings were excused, to some extent, by the fact that their times were not auspicious. Still, Obama is fairly clear that if the party is to move forward it must return to the standards set by earlier exemplars, and especially by its heroes who brought about major political changes lasting for a generation or more. This was the context of his comparison of Clinton to Ronald Reagan, which raised such a ruckus early in the campaign:

> I do think that, for example, the 1980 election was different. I think Ronald Reagan changed the trajectory of America in a way that Richard Nixon did not and in a way that Bill Clinton did not. He put us on a fundamentally different path because the country was ready for it.[14]

The comparison of Clinton to Nixon is delicious in its own right, but Obama's larger point is that Clinton was no Reagan, partly because the times were different but mostly, as he points out in his book, because Clinton was undisciplined, unfocused, and conceded too easily to the Right. Once his own health care reform—Hillarycare, as it came to be called—had been quashed by a Democratic Congress, Clinton decided the Age of Reagan was here to stay, and gave up on his already halfhearted efforts at political realignment.

Obama assumes the Reagan Revolution is not here to stay, because the Obama Revolution is just beginning. As tokens of Obama's seriousness about fundamental political change, *The Audacity of Hope* mentions Franklin D. Roosevelt more often than it does any living Democratic politician. And it features a long, interesting discussion of the Declaration of Independence and the Constitution, the political point of which is to reassert the Democrats' claim to speak for American ideals, which are the touchstone of every electoral realignment.

Thus the commentators who interpreted Obama as a new kind of postpartisan political figure get it exactly wrong. It's true he wants to stop "arguing about the same ole stuff," as he once put it; he wants to move beyond the decades-long debate between liberalism and conservatism. Bill Clinton attempted the same thing in 1992, as did George W. Bush in 2000. The forty-second and forty-third presidents each promoted a novel doctrine he hoped would leave the old arguments in the dust—the Third Way and compassionate conservatism, respectively—but each proffer eventually fizzled. Though capable of distracting the pundits for one or two elections, neither idea came close to new-modeling the public philosophy and realigning the entire party system. But Obama, as political scientist James W. Ceaser noted, dispensed even with the obligatory bow to such a new doctrine.[15] Instead of a fresh public philosophy, he offered himself. John McCain reciprocated. As a result, the 2008 presidential race took place squarely within the familiar ideological framework of liberalism and conservatism, but with McCain promising some maverick departures from, while still accepting, the right-wing norm, and Obama pressing for leftward hope and change. Once in office, Obama tried belatedly to christen his reforms "the New Foundation," but abandoned the name amid a torrent of late-night TV jokes about ladies' undergarments. Retreating from any suggestion of newfangled ideology, he then came up with a second, even feebler mantra, "Win the Future," which in its desperation recalled President Gerald Ford's mid-1970s rallying cry, "Whip Inflation Now," abbreviated on political buttons as "WIN." A wag asked why couldn't the message be "Stop Inflation Now," or "SIN"?

Yet a change is gonna come, Obama vows, even if it won't come in the form of a new doctrine. He doesn't need a new ideology because he finds plenty of life left in the old, a concession that in no way detracts from his ambitiousness. For the change he seeks is nothing less than an electoral earthquake that would permanently shatter the 50-50 America of the past four presidential elections (prior to 2008). He thinks liberals can get beyond the old debate by finally *winning* it. "Eking out a bare Democratic majority isn't good enough," he wrote in *The Audacity of Hope*. "What's needed is a broad majority of Americans—Democrats, Republicans, and independents of good will. . . ." Though he didn't call explicitly for a realignment that would make the Democrats the majority party again, he left little doubt which party would be the centerpiece of the new coalition. After the New Hampshire primary, he told his supporters "you can be the new majority who can lead this nation out of a long political darkness." A month later, after winning the Wisconsin primary, he explained what he called "my central premise," that "the only way we will bring about real change in America is if we can bring new people into the process, if we can attract young people, if we can attract independents, if we can stop fighting with Republicans and try to bring some over to our side. I want to form a working majority for change." Speaking to the AFL-CIO in 2003, he laid out the long march that would be necessary:

> I happen to be a proponent of a single-payer universal health care program . . . [a] single-payer health care plan, a universal health care plan. And that's what I'd like to see. But as all of you know, we may not get there immediately. Because first we have to take back the White House, we have to take back the Senate, and we have to take back the House.

As a presidential candidate, he was not officially for a single-payer health care plan; his 2008 platform stopped far short of that. Nor did he ever repeat this candid statement of his ultimate goal, which could be described by that hoary but accurate curse word, *socialized medicine*. In the meantime, however, the Democrats in 2006 handily recaptured both the Senate and the House of Representatives, followed by the presidency in 2008. The "working majority for change," or a reasonable facsimile of it, was ready to get to work.

## Lightworker

If the leading edge of Obama's audacity is his desire to bring about fundamental political change at a time when he thinks every other leading Democrat has given up on it or lacks the gifts to achieve it, his daring shows itself too in his confidence that he is "the one," as Oprah Winfrey famously said. His supreme self-confidence bordering on self-obsession isn't news, of course, either to his critics or to his admirers. They gush over him less than they used to, but it's good to be reminded of their initial enthusiasm, their enthrallment. Here's Mark Morford, an online columnist in San Francisco, in 2008:

> Many spiritually advanced people I know (not cower-
> ingly religious, mind you, but deeply spiritual) identify
> Obama as a Lightworker, that rare kind of attuned being
> who has the ability to lead us not merely to new foreign
> policies or health care plans . . . but who can actually help
> usher in *a new way of being on the planet*, of relating and
> connecting and engaging with this bizarre earthly experi-
> ment. These kinds of people actually help us *evolve*. They
> are philosophers and peacemakers of a very high order, and
> they speak not just to reason or emotion, but to the soul.

Yet the fact is that neither then nor now has the precise character of Obama's soulcraft been well understood. In his own terms, he seeks to bring about enduring political change even as (to mention those he often invokes in this connection) Ronald Reagan, Franklin Roosevelt, and Abraham Lincoln did before him.[16]

Obama's account of Lincoln deserves particular attention. Lincolnian language appears and reappears in Obama's speeches, usually in dreadful, tin-eared paraphrase. He often compares himself indirectly and sometimes directly to the first Republican president.

As he prepared to take office, he read and encouraged reporters to read *Team of Rivals*, Doris Kearns Goodwin's book about Lincoln's cabinet. On his way to Washington, he retraced part of Lincoln's journey from Springfield, Illinois, in 1861. At his inauguration, he took the oath of office with one hand on Lincoln's Bible. At the luncheon afterward, each course had some reference to the sixteenth president and was served on replicas of Lincoln's china. The speech that initially put Obama on the map, his 2002 denunciation of the pending Iraq War, concludes in a mangle of the Gettysburg Address: "Nor should we allow those who would march off and pay the ultimate sacrifice, who would prove the full measure of devotion with their blood, to make such an awful sacrifice in vain." He couldn't have had a speechwriter for that one. He announced his candidacy in Springfield, a place central to Lincoln's political career and site of some of his great speeches, including the "House Divided" and his affecting farewell to the city as he left to assume the presidency. In his speech, Obama does his best to appropriate Lincoln's memory:

> And that is why, in the shadow of the Old State Capitol, where Lincoln once called on a divided house to stand together . . . I stand before you today to announce my candidacy for President. . . . By ourselves, this change will not happen. Divided, we are bound to fail. But the life of a tall, gangly, self-made Springfield lawyer tells us that a different future is possible. He tells us that there is power in words . . . in conviction. That beneath all the differences of race and region, faith and station, we are one people. He tells us that there is power in hope. As Lincoln organized the forces arrayed against slavery, he was heard to say: "Of strange, discordant, and even hostile elements, we gathered from the four winds, and formed and fought the battle

through." That is our purpose here today. That's why I'm
in this race. Not just to hold an office, but to gather with
you to transform a nation. . . . Together, starting today,
let us finish the work that needs to be done, and usher in a
new birth of freedom on this Earth.

Obama strains to let you know he identifies, as we say today,
with Lincoln: Abe is not the only "tall, gangly, self-made" lawyer
primed for greatness that the audience is supposed to recognize.
Though it ends with another paraphrase of the Gettysburg Address, the passage—and the whole speech—is meant to recall Lincoln's "House Divided" speech, which kicked off his campaign for
a U.S. Senate seat in 1858 against Stephen Douglas. The quotation
about the "four winds" is Lincoln's description of the new Republican Party, forged from fragments of the fading Whig and Free Soil
parties, and reaching out to antislavery Democrats and centrists.

Thus Obama compares the new majority he seeks to build to
the majority party that Lincoln helped create. He tries to inspire
Democrats by appealing to the founder of the generations-long,
post–Civil War Republican majority. This is partisan ambition of
a high order, masquerading as high-toned bipartisanship or postpartisanship. Obama speaks as though Lincoln had been trying to
overcome the country's divisions by calling for unity, for fraternal
or bipartisan or postpartisan cooperation in the spirit of national
renewal. In fact, Lincoln warned in Springfield that the Union
would "become all one thing, or all the other." It would become
either all free, or all slave. Lincoln's road to unity (for of course
he did wish to save the Union and free government) ran through
division, through forcing the country to choose: slavery or freedom. That's why it's called the "House Divided" speech. Lincoln
offered no compromise on the issue of slavery's moral status or its
extension into the western territories. In fact, he was determined to
undercut the false compromise then being peddled by Douglas in

the form of popular sovereignty. Obama's political point seems to be similar, despite his soothing and disingenuous language: all our chronic divisions will be healed once the country is safely in the hands of a broad, liberal, pro-change majority. That is his "central premise." *Then* postpartisanship will indeed flower, once partisanship is laid to rest by the effective disappearance of conservatives after a shattering Republican defeat.

## The Audacity of Hope

Obama spoke in 2005 at the opening of the Lincoln Presidential Library and Museum in Springfield. On that occasion, he talked more of the man himself than of his political legacy. Lincoln exhibits "a fundamental element of the American character," he said, "a belief that we can constantly remake ourselves to fit our larger dreams." He hailed Lincoln's "repeated acts of self-creation, the insistence that . . . we can recast the wilderness of the American landscape and the American heart into something better, something finer." The wilderness of the American heart—now that's a phrase Lincoln would never have uttered. Is it Obama's own view of the American soul's desolation? Perhaps it is even a topic of family conversation. Michelle Obama used to say in her standard 2008 campaign speech that the country is "just downright mean" and "guided by fear," though she implied the problem was curable. Her husband "knows that at some level there's a hole in our souls," she often said, and he "is the only person in this race who understands that before we can work on the problems, we have to fix our souls. Our souls are broken in this nation." In a speech in Los Angeles she amplified the point: "Barack Obama . . . is going to demand that you shed your cynicism. . . . That you come out of your isolation, that you move out of your comfort zones. That you push yourselves to be better. And that you engage. Barack will never allow you to go back to your lives as usual, uninvolved, uninformed."[17]

Obama calls this power of soul-fixing "the audacity of hope."

The phrase comes from a sermon by the man who married Barack and Michelle, baptized their daughters, and served as their minister for twenty years, the Reverend Jeremiah Wright. Obama defines it as "a belief in things not seen," applying St. Paul's description of Christian faith to the earthly transformations promised, consistently and seriously, by the president and by American liberalism for more than a century. He isn't the first or the last Lightworker, after all, though in some respects he is a new political phenomenon. To state the obvious, he is young, gifted, and black, and as Nina Simone (followed by Aretha Franklin) sang forty years ago, "To be young, gifted, and black / Is where it's at!" Even after the past few years' disappointments, most Americans feel a certain pride in Obama and his achievement. Beyond that, his rare combination of Ivy League degrees and Chicago street cred, of high-sounding post-partisanship and hard-core partisanship, leaves people guessing. To call this combination or alternation "pragmatic," as he likes to, is simply to accept his invitation not to think about it.

But in the decisive respect, Obama does *not* stand for something new under the sun. He represents a rejuvenated version of the progressive impulses that gave birth, a century ago, to modern American liberalism. Most political movements in our history came into being to press a putative reform or two, and dissolved when they had succeeded, or failed, definitively. The antislavery and women's suffrage movements succeeded, the Populists' demand for the free coinage of silver failed, and Prohibition is an interesting case of a movement that succeeded and *then* failed. Modern liberalism is something else again.

What came to be called liberalism was the first intellectual and political movement without a clearly defined goal of reform, without a *terminus ad quem*: the first to propose an endless future of continual reform. It pledged to make the country "progressive," to keep America always moving forward, up-to-date, and in tune with the times. No specific reform or set of them could satisfy that demand, and no ultimate goal could comprehend, much less specify, all the changes in political forms and policies that might become necessary in the future. Obama's campaign slogans in 2008 were marvelous examples of this open-endedness. Hazy as they were, they galvanized millions of voters in the primaries and general election, and judging by his record as president they are likely to remain his most renowned utterances. I mean the famous monosyllables, *hope* and *change*.

Around these two gas giants Obama's rhetorical system spun. To begin with, in Obamaspeak hope confronts "the politics of cynicism." Cynicism wears many hats in Obama's speeches: the "politics of anything goes," the tactics of "spin masters" and "negative ad peddlers" who seek to divide us, the viewpoint of those who think "politics has become a business and not a mission," the "can't-do, won't-do, won't-even-try style of politics," the resort to "stale tactics to scare the voters," the effort to "make a big election about small things." The nefarious cynics deny the nation's problems, then blame them on someone else ("the other party, or gay people, or immigrants"). And when the people look away "in disillusionment and frustration," the "cynics, and the lobbyists, and the special interests" move in to fill the void, and turn government "into a game only they can afford to play."[18]

Cynicism remains Obama's all-purpose explanation for the "gridlock and polarization" that characterize our small politics today, that rob Americans of "our sense of common purpose—our sense of higher purpose." He has disdain enough for Democrats as well as Republicans, but lately he contemns primarily the alleged extremism of the congressional Republicans, their stubborn refusal to compromise. Since he finds it difficult to believe that he's not being reasonable, and that Republicans and Tea Partiers could possibly believe, on the merits, what they claim to believe, he tends to resort sooner or later to the notion that his opponents must be rationalizing. That is, they have interests or prejudices to defend that they couldn't possibly admit publicly. To reason or deliberate with them, at least in the high-minded sense, is therefore futile. Even self-styled rebels like the Tea Partiers are hiding their real motivations, often from themselves. Their various "absolutisms"—of the free market, Christianity, and majoritarianism—are the flip side of their own private desperation. Their leaders cynically exploit the rank and file, who though not conscious cynics themselves are, typically, victims of powerful forces outside their comprehension or control. They "cling," therefore, as Obama said once in San Francisco, "to guns or religion or antipathy to people who aren't like them," out of anxiety over their falling social status or declining job prospects. "A cynical electorate is a self-centered electorate," he argues, even if they don't realize it. In any event, the bogeyman of cynicism is his equivalent of the deeper and richer institutional analysis that once, as we shall see, was American liberalism's stock-in-trade. Liberals used to criticize institutions like the congressional committee system, the separation of powers, and political parties dominated by large donors and local interests, on the theory that they perverted American politics by marrying narrow self-interest to mechanistic checks and balances. The result, supposedly, was the obstruction of salutary change and progressive reform—the dreaded "deadlock of democracy." Obama in 2008 simplified this elaborate protest literature into an indictment of the Tom DeLays and Karl Roves of contemporary Washington who had induced public apathy, a paralyzing loss of faith in the future, and a dumbing down of political debate. He clings to that simplification today, though he's had to broaden the villain class.[19]

The antidote for widespread cynicism and rationalization is *hope*, the saving quality that a Lincoln or FDR can awaken in the public, at long last. Without that change in the public mind or soul, no broad program of legislative reform can be enacted; and without such reforms, the belief in hope cannot long be sustained. "We know the challenges. . . . We've talked about them for years," Obama argues.

What's stopped us is the failure of leadership, the
smallness of our politics—the ease with which we're dis-
tracted by the petty and trivial, our chronic avoidance of
tough decisions, our preference for scoring cheap political points instead of rolling up our sleeves and building a
working consensus to tackle big problems.

There are good "ten-point plans" aplenty, he observes. They
have all been blocked by the pervasive cynicism in Washington,
and concerning Washington, over the past three decades or more.
"It's time to turn the page," he likes to say, but the problem is that
"we haven't had leaders who can inspire the American people
to rally behind a common purpose and a higher purpose." The
public has to be roused from its cynical slumbers, and though
he admits that the times must be right, it's up to the leader to
recognize and exploit the favorable moment, to open the public's
eyes to the possibility, indeed the imminence, of a political leap
forward.[20]

In every critical moment, Obama affirms, "a new generation
has risen up and done what's needed to be done. Today . . . it is
time for our generation to answer that call." Echoing the passing of the torch that JFK evoked on behalf of his generation—as
well as the heroic exertions of the civil rights movement—Obama
in announcing his candidacy adjured his followers, "Let's be the
generation that ends poverty in America. . . . Let's be the generation that finally tackles our health care crisis. . . . Let's be the
generation that finally frees America from the tyranny of oil." He
continued:

That is why this campaign can't only be about me. It
must be about us—it must be about what we can do together. This campaign must be the occasion, the vehicle, of
your hopes, and your dreams.

His oft-repeated refrain about "the moment" speaks to this generational awareness. The moment is when a generation becomes conscious of itself and ready to act to change America, when it realizes "yes, we can" change the political system. This awakening that makes possible all subsequent political reform is what Obama refers to when he proclaims, "We are the ones we've been waiting for. We are the change that we seek." The "moment" is when hope vanquishes cynicism, when the sudden realization of generational unity finally makes hope, and thus liberalism itself, viable. The crucial change is what social scientists in another context used to call "the change to change," the point at which a people inured to stasis and stalemate begins to expect instead continual improvement. Obama invites new voters, college students, independents, and disaffected Republicans to join in this new political revelation.[21]

The clash between hope and cynicism pits the future against the past, "a new politics for a new time" against the same old, same old. He is careful to distinguish hope from blind optimism. Hope is not "the almost willful ignorance that thinks . . . the health care crisis will solve itself if we just ignore it." "Hope is not ignoring . . . the challenges that stand between you and your dreams." On the contrary, hope is "imagining, and then fighting for, and then working for, struggling for what did not seem possible before."

> You know, there is a moment in the life of every generation when that spirit has to come . . . for what we know in our gut is possible . . . when we determine that we're going to keep the dream alive for those who still hunger for opportunity and still thirst for justice.

That's why he insists on "the audacity of hope," which he calls "God's greatest gift to us, the bedrock of this nation; a belief in things not seen; a belief that there are better days ahead."[22]

This mixture of divine and secular assertions boils down to a moment of historical clarity: *hope* is when you recover your faith in progress, and so in progressivism; when you can envision "the better days ahead," but also when you resolve not to wait passively for them but to pursue them courageously: the *audacity* of hope. To the leader belongs the awesome power to transmit this vision of the future to the people, to allow those who walked in darkness to see a great light, and to organize them for the "march into the future" that he alone can lead. Despite Obama's efforts to offload to or share some of this responsibility with the awakened people, it falls primarily on his head as the prophet of the moment. Though he expresses doubts about his ability and invites the people to correct him when he goes wrong, he knows he is the one. It's like the old joke in which the egotist says, "But enough about me. Let's talk about you. What do you think about me?" That's why Obama's campaign catchwords lack grammatical subjects and objects. Who should change, and in what way? Hope for what, exactly? Yes, we can . . . do what again? These slogans discourage deliberation. They hover childlike, dreamlike over all debate. Each requires some external agent to supply and define its missing terms. Together they say, in effect, give us a leader who will show us what to hope for, what to change, and what we can do.

*Audacity* is a curious word with two meanings, which reflect a genuine moral ambiguity. It means both boldness, daring, confidence—and reckless daring, rashness, foolhardiness. It can be a good or a bad thing, a virtue or a vice. "Hope," by contrast, is a passion; in the language of the medieval schools, hope aims at a future, arduous, and possible good. It doesn't always attain that good, but young people tend not to know that; their inexperience of life and of their own shortcomings disposes them to be full of hope and heedless of dangers, noted Aristotle, just as drunks are. Obama, like most liberals, looks to the young for confirmation of his hopes and dreams, and urges the old to adopt the inexperienced and immoderate as their guide. Hope is also a theological virtue, but presumably even Obama doesn't mean to offer eternal happiness to his followers.[23] His vision is of earthly happiness, wholeness, and justice, a world in which all things desirable are possible. As he helpfully explained in his debut at the 2004 Democratic National Convention, his name, Barack, means "blessed." It's as though he were a little bit of heaven on earth, intent on helping turn earth into heaven, even if he wasn't born in a manger.

As he told Congress in the late innings of the health care fight, "we did not come here just to clean up crises," even one as big as the Great Recession. "We came to build a future," to do the "great things" that "will meet history's test." He concluded, "This is our calling. This is our character." And that had been his ambition all along. "Let us transform this nation," he implored in 2007 when he announced his candidacy for president. After the 2008 Iowa caucuses, he hailed those voters "who have the courage to remake the world as it should be," and after the New Hampshire caucuses he declared, "Yes, we can repair this world. Yes, we can." As Election Day approached, he promised, "We are five days away from fundamentally transforming the United States of America."[24]

Those words mean *this will be a different country* when he's finished with it. If, Rip Van Winkle–style, one had slept through the Obama administration, one would awaken, as it were, in a new land. The old word for such a profound change was *revolution*. As a good progressive, however, he reckons his revolution will be one in a series, an unending series generated by social progress or history itself. His reforms will connect to Wilson's, FDR's, and LBJ's before him, and others yet to come, and all these together will constitute a continual upward evolution. That sounds reassuring, insofar as it promises to take the sting and surprise out of change; but such inevitability comes at the expense of liberty, because there is no choice about the whole of liberal-style progress. In the old days, one could choose to make a revolution or not. A revolution could be defeated or reversed. But you cannot deliberate about the inevitable, which is how progressives think of history. As we've been told for generations now, ad nauseam: you can't turn back the clock.

## Heightening the Contradictions

By the same token, however, you can't turn the clock ahead, either. What Obama invokes as "history's test" is a stern one: success or irrelevance, power or nothingness, to recur to Tomasky's suggestive language. Either you're on the right side of history or the wrong side, where the right side is necessarily understood to mean the winning side, and the wrong side the losing one. Otherwise this would not be a *historical* test but an abstract moral or philosophical one. The obvious moral difficulty—does the right side always win its wars?—can be finessed for a while by distinguishing between wars and mere battles. It's possible to lose many battles and still win the war, eventually. But "history's test" is of necessity a final examination; it can't be postponed indefinitely without the whole idea of historical validation becoming a laughingstock or an otherworldly stalking horse, neither of which liberalism fancies itself to be. Meeting history's test, as Obama sees it, means recognizing that the "moment" has come for bold, new reforms; but if these prove untimely and unattainable, if the moment comes and goes fruitlessly, then it casts doubt not only on the prophecy and the prophet but on the whole prophecy business. If Obama cannot repeal the George W. Bush–era tax cuts, if he cannot close the Guantánamo detention facility, if he cannot get the debt ceiling raised without agreeing to federal spending cuts, those are battles lost. If he cannot get reelected, that's a defeat of an altogether more serious sort. If the new majority for change does not triumph in congressional and state elections in 2012 or 2014 or 2016, then the long-term prospects for a liberal consummation drop still further. If Obamacare is repealed and replaced after 2012 by an energized conservative majority that controls the presidency, Senate, the House of Representatives, and most state legislative chambers and governorships, then Obama's legacy and his claim to leadership will lie in ruins.

Even so, American liberals would try to overcome their embarrassment by insisting that poor Obama was too far ahead of his time. Desperate as it is, that argument is neither unprecedented nor implausible, and it has the capital advantage of being unfalsifiable. But it would certainly be a stretch, because it would highlight, by trying to ignore, the dispiriting truth that Obama had it *won*— had Obamacare enacted and written into law, its implementation under way—only to suffer the ignominy of defeat. After the repeal of Prohibition, for example, how many observers concluded that the problem with the Eighteenth Amendment was that it had been ahead of its time? After the dissolution of the USSR, how many Russians, or even communists, defended the extinct Soviet Union as too good for this world, or tragically in advance of its age? It's one thing to claim grandiloquently to represent the future, to *be* the future, ever glorious and ever distant. It's quite another to *have been* the future. The former trades in utopian speculation, however scientific the speculation claims to be. The latter forces one, wearily, to confront a history of failure and disillusionment—to confess "the god that failed," to borrow that ever resonant term from the Cold War.

American progressives' favorite tense is future perfect; they hate like hell to wrestle with past imperfect. So President Obama faces, by his own standards, a crucial test of his leadership. The election of 2008 proved, as that of 1992 had as well, that post-Reagan Democrats could win control of all three elected branches of the national government. In his first two years in office Obama further demonstrated that the Reagan legacy, both ideological and institutional, had not rendered impracticable an aggressive agenda of liberal social reform and government expansion. Now he faces an electorate that in 2010 moved dramatically rightward, even as his policies were moving American government briskly leftward. Among other things, he has to show both liberals and conservatives that the future is on *his* side, not Reagan's, and that the voters will come around to his "new politics for a new time." Tacking rightward, as Bill Clinton did in the 1990s, is out of the question, because Obama's whole project, though he would never put it so candidly, is to prove that the era of big government is *not* over. Whatever rhetorical and even policy concessions he may feel compelled to make, you can be sure they will be minor compared to Clinton's. Obama has in effect doubled down on the Left's bet on big government, and it is too late to take the chips off the table now.

But his bet comes just when the political economy of the welfare state is reaching a turning point, both in the United States and Europe. Everyone knew, vaguely, that with baby boomers beginning to collect benefits and fewer young workers available to pay taxes, the welfare state would hit a demographic wall eventually—a decade or two or three down the road. That crisis, of unfunded liabilities and revenue shortfalls, is still to come, in fact. The current crisis is related, but different in origin. The wall that Europe is hitting, and that we are coming up on fast, is a wall of deficits and debt. Although unrestrained entitlement spending is a part of it, the immediate problem was precipitated by the financial crisis of 2007–2008, the ensuing Great Recession of 2008–2009, and governments' reaction to these shocks. Rather than learn from the Bush administration's fiscal and monetary mistakes, the Obama administration compounded them. Obama's costly stimulus and bailouts stimulated mainly the deficit, and the Federal Reserve flooded the economy with money to prevent the banking system's collapse and to prop up economic growth. Unemployment insurance and other "automatic stabilizers" cost much more than anticipated, tax revenues dropped due to the prolonged downturn, and rather than cut discretionary spending the administration piled on more. In other words, the same rock-bottom interest rates, massive deficits, and credit-fueled consumption boom that helped get us into the financial crisis are now being counted on, by a bit of Keynesian magic, to get us out of it.[25] When the banks couldn't handle their bad debts, they sent them to the U.S. government. But to whom does the government turn when it can't pay its bills?

Even as Greece proved insolvent and many other European countries teetered on the brink, Obama's policies on health care, taxation, and regulation pushed America further toward the European model of social democracy. In effect, his audacity made the problems of the American welfare state worse and more urgent. His policies made the chronic inability of big government to "make payroll," as William Voegeli dubs it, that much more acute.[26] The advent of the true crisis of the welfare state has been accelerated, the hole into which it will plunge the economy dug deeper, and the options for dealing with the chronic shortfalls made worse. The Marxists call this policy of speeding up the social and political reckoning "heightening the contradictions." It's possible that Obama wants to heighten the contradictions in order to bring about a crisis of the American welfare state that would be solved by its engorging another 10 percent or 20 percent of the American economy: the Swedenization of America. Perhaps, though, he is content to win the moral battle—a historic expansion of the welfare and regulatory state—and leave it to the next administration to wage the fiscal one. Or maybe he believes his own talking points, and regards Obamacare, green energy, and Dodd-Frank as reforms that will save money over the long run. But whatever his intention, even he acknowledges now, with economic growth in the doldrums, that the welfare state cannot continue indefinitely along the same paths.

What history confronts him with, in short, is not merely a test of his own leadership but also a test of liberalism's credibility as the once-and-future American public philosophy. More and more, the blue-state social model, as Walter Russell Mead calls it, looks anachronistic and unimaginative—behind rather than ahead of the times. A health care reform bill, to take the central example, that stretches to three thousand pages and creates 159 or so (an exact accounting proves slippery) new boards, commissions, and agencies hardly betrays the nimbleness, efficiency, transparency, reliability, and personalization that Americans expect from new companies, products, and services at their best. Liberalism seems about to succumb to the very critique it once leveled disdainfully at the old American constitutional and political order: the failure to evolve. Beyond its bureaucratic shortcomings, however, looms a deeper problem with liberalism's understanding of human nature and the purposes of government, which led it to presume to lead and administer a free society and concoct rights to health care, housing, and a job in the first place. Heightening the contradictions could soon produce a kind of revolution all right, but not the one President Obama believes in and anticipates.

# 2

## Woodrow Wilson and the Politics of Progress

**D**espite their extravagance, the phrases come easily to Barack Obama's lips: let's "meet history's test" by "fundamentally transforming the United States of America." Why stop there? While we're at it, we can repair the moral universe by decisively closing the gap between the real and the ideal. Seizing the moment, and with our best change-agents in charge, we can "remake the world as it should be."[1] This trust in "fundamental" but never final transformations, in continual progress toward an unspecified but ever more egalitarian condition of social justice and political wholeness, inspired and guided by visionary and compassionate leaders, themselves inspired and guided by history with a capital *H*, and the entire cosmic process culminating in the growth of the State with its master class of expert administrators—this is modern liberalism in a nutshell. It began almost exactly a century ago in the Progressive movement, determined to reform and re-form American political life. During the New Deal it changed its name and some of its methods but not its essential goals. Obama's asseverations come easily to him, in part then, because statesmen whom he admires have asserted them many times before. They sound familiar because they *are* familiar, the topoi of liberal rhetoric that have been drawn on by generations of journalists, professors, and politicians.

It wasn't always so. In 1886, the year after the appearance of his first book, *Congressional Government*, and while he was toiling away as a young faculty member at Bryn Mawr, Woodrow Wilson wrote to a close friend:

> I believe . . . that if a band of young fellows (say ten or twelve) could get together . . . upon a common platform . . . with reference to the questions of the immediate future, [and] should raise a united voice in such periodicals, great or small, as they could gain access to, gradually working their way out, by means of a real understanding of the questions they handled, to a position of prominence and acknowledged authority in the public prints, and so in the public mind, a long step would have been taken towards the formation of such a new political sentiment and party as the country stands in such pressing need of—and I am ambitious that we should have a hand in forming such a group. All the country needs is a new and sincere body of thought in politics, coherently, distinctly, and boldly uttered by men who are sure of their ground.

Three years earlier he had disclosed to his fiancée a "solemn covenant" he had made with the same friend, Charles Talcott, while both were undergraduates at the College of New Jersey (as Princeton was still called then). "We swore," Wilson said,

> that we would school all our powers and passions for the work of establishing the principles we held in common; that we would acquire knowledge that we might have power; and that we would drill ourselves in all the arts of

persuasion, but especially in oratory . . . that we might have facility in leading others into our ways of thinking and enlisting them in our purposes.

Not your typical student (nor faculty) braggadocio, but then from the beginning Wilson was keen to win a new kind of political greatness for himself. He set out to found, or help found, "a new political sentiment and party" that would appeal to the "thirst & enthusiasm for better, higher, more hopeful purpose in politics than either of the present, moribund parties can give."[2] Beginning as an intellectual insurgency in universities and high-toned journals, but always intent on assuming political power behind leaders with well-educated ambition and well-honed oratorical skills, this was liberalism in embryo, the new movement for which Wilson aspired to conquer first the public mind, and then the public.

He was not the only young man to feel the tug of these new ideas. But he did something that none of his other great contemporaries, not even Theodore Roosevelt, managed or quite attempted. As undergraduate, graduate student, professor, and university president, Wilson spent three decades in the academy contemplating the failings of the old American constitutional system, testing his critique of it, and preparing the rhetorical case for its transformation. He had developed a new theory of the presidency and of the whole political system long before becoming president; no president since has done anything remotely comparable, and none before him, either, except the founder-presidents John Adams, Thomas Jefferson, and especially James Madison. But though they'd thought deeply and written much about the overall design of the political system, none of these famous thinkers had conceived of the executive as the center and leader of it, as had Wilson. To be sure, Alexander Hamilton had insisted on the necessity of an energetic executive in American politics—but he never became president, and besides, "energy" was not the quality Wilson was looking for in the position, much less the constitutional fidelity on which Hamilton had insisted even in his most high-flying moments.[3] Among Wilson's immediate predecessors and his opponents in 1912, William Howard Taft was deeply learned in the law, a cagey and effective administrator of the Philippines, and an astute defender of representative government; but his political thinking always had a touch of Whiggish legalism to it. In his wide-ranging career his favorite job was chief justice of the U.S. Supreme Court. Teddy Roosevelt certainly broke new ground in office but vouchsafed his "stewardship" theory of the presidency only after the fact; at the time, to many observers, his guide seemed to be mostly animal spirits. As his chief opponent in 1912 noted, acidulously, "life does not consist of eternally running to a fire."[4]

Of the great captains of Progressivism, Woodrow Wilson was the most clear-sighted and the most consequential. Not only one of its supreme theorists, he stole from TR the honor of being the principal model for future progressive political striving in general, and especially for the inspirational parts of presidential leadership. As Richard Hofstadter was not the first to point out—TR whaled away at the theme—Wilson was in some respects a very conservative Progressive.[5] In 1912 he sought to position himself as the moderate, the all-round sensible reformer, with Taft the stand-patter to his right and the increasingly radical and desperate Roosevelt to his left. It worked, though the salient fact was that Taft and TR split the GOP vote, allowing Wilson with 42 percent of the popular vote to become only the second Democratic president since the Civil War. Far from being disqualifying, Wilson's "conservatism" was the flip side of his incipient liberalism. The roots of his peculiar conservatism were twofold, based on book learning at Princeton and later Johns Hopkins, and on his own political bloodlines, as it were. His family was southern by choice, having moved from Ohio to Staunton, Virginia, where Thomas Woodrow was born, and then to Georgia; as a young boy he had seen Robert E. Lee travel through Atlanta after the surrender. The Democratic Party was rooted in the Solid South in Wilson's lifetime, and he claimed to understand the South instinctively, especially its celebration of custom and family life, and its suspicion of individualism and "abstract" reason and equality. His Progressivism would prove remarkably congenial to southern prejudices—as president, he would resegregate the District of Columbia and drive from office many black federal employees—though there was never any doubt he wished to subsume southern-style conservatism into Progressivism rather than vice versa.[6] He saw the affinities with the antebellum and postbellum South that the Republican Progressives like TR did not, or at least disdained. His youthful desire to found a new political party thus easily gave way to the desire to instill a new sentiment into the old Democracy. Wilson helped to begin the conversion of the states' rights, reactionary Democrats into the national party of liberalism. FDR completed the conversion (the Dixiecrat influence lingered for decades, to be sure), but Wilson was among the first to conceive that the losing party in the Civil War, stained by its support for slavery and disunion, could be reborn as the hopeful and egalitarian party of the future; that the champion of the sovereign states could reinvent itself as the champion of the progressive State. Wilson's so-called conservatism thus helped to make him the founding father of Democratic Party liberalism.

## The Progressive Movement

Progressivism was much bigger than Woodrow Wilson, of course. As an intellectual then political movement, it was eclectic, so eclectic that some historians beginning in the 1970s pronounced it a will-o'-the-wisp.[7] But no one at the time thought Progressivism so various and contradictory as to be meaningless, much less nonexistent, though its adherents battled furiously over its political agenda. In 1912, each of the three major presidential candidates considered himself a Progressive, though only Theodore Roosevelt ran as the candidate of the Progressive Party. The raging disputes among TR, Wilson, and Taft showed the difficulty of reducing a complex intellectual movement to a single antitrust policy, or a common attitude toward the initiative, referendum, and recall. But even bitter disagreements over policies did not obscure the premises shared by almost every major figure who called himself or herself a Progressive. The most intense disputes usually occur among coreligionists, after all. Progressivism's intellectual roots go back to the 1890s and even earlier—to the rise of the American research university after the Civil War, and to the shifting focus of politics after Reconstruction. Historians, journalists, and ordinary citizens understood that at the time, and most historians since, even those addled by the 1960s, have admitted the obvious, too.[8]

As a political label, "the Progressive movement" did not come into widespread use until after Teddy Roosevelt's presidency, which ended in 1909. The term peaked during his double-barreled run for the same office in 1912, first in the Republican Party and then, when he realized he couldn't win the nomination and bolted the GOP national convention, as the alpha moose of the Progressive Party. But as a term for a certain reformist sensibility and an emerging set of issues, the adjective *progressive* had been current since the 1890s—first in England and then in Germany and America. By the time Herbert Croly and Walter Lippmann launched the *New Republic* in 1914, they naturally embraced the term to define their magazine's political position, and by the same token proudly enrolled themselves in the crusading Progressive movement.[9]

Most Progressives assumed that the social problems and grievances they confronted were new, outgrowths of the unprecedented development of the modern economy, and proof that one stage of American politics had ended and another, more difficult, and less clearly defined one had begun. "There are no precedents to guide us," lamented Lippmann, "no wisdom that wasn't made for a simpler age."[10] The old politics had ended with the Civil War, not in a blaze of dialectical clarity but amid the smoke and noise of battle. Arms had not yielded to the toga, a point not lost on Oliver Wendell Holmes Jr., and many other future Progressives who fought in the war. For four score and nine years American politics had revolved around the nature ("partly federal and partly national," Madison said cryptically) and expansion of the Union, the status of slavery within it, and the resulting powers of the national government. But as the Progressives saw it, those issues had been settled, once and for all, by Appomattox: the Union would brook neither secession nor slavery, and the Constitution would furnish whatever powers were necessary to the Union's preservation. The primacy of the nation had been established in fact, whatever its status in theory might have been. Clarified and strengthened by the Civil War amendments (the Thirteenth, Fourteenth, and Fifteenth), the Constitution stood confirmed now as an antislavery document. Law and will and right were one.

Actually, there was a great deal of unfinished business left over from the Civil War, having to do especially with the civil and voting rights of black Americans. When Reconstruction ended and the last battalions of federal troops were withdrawn from the South in 1877, those issues were left in the hands of southern whites, who began almost immediately to erect the legal superstructure of Jim Crow. One might think that for Americans keen on crafting a purer and higher politics, here lay a vast open field for their best efforts. Whatever efforts the would-be Progressives devoted to the vindication of black citizens' rights, however, were certainly not their best, nor even very numerous. Though no friend of slavery per se, Wilson was incensed by Reconstruction and often defended or at least extenuated the South's resistance to it. In his five-volume history of America, for example, he wrote, "the white men of the South were aroused by the mere instinct of self-preservation to rid themselves, by fair means or foul, of the intolerable burden of governments sustained by the votes of ignorant negroes and conducted in the interest of adventurers."[11] The nascent Progressives weren't the only ones to lose interest in the unfinished Reconstruction agenda; the whole country was straining to reunite the sections, to put the issues of the war behind it, to face the future like a normal nation. Still, the belief that history obeyed the laws of progress provided an additional moral sedative, calming fears that a discriminatory regime in the South and elsewhere could be other than temporary and, in the end, even beneficial for blacks. For most Progressives there was no need to focus on the fading issues of a fading—and simpler—era, which would take care of themselves in due course. The question was, what were the issues of the dawning age? What was American politics supposed to be about henceforth?

The millions who enlisted in the Progressive movement, small businessmen, newspaper editors, teachers, suffragettes, lawyers, doctors, ministers, and many others, worried about the new conditions of life facing the country—daunting problems like mass immigration, urban slums, declining economic opportunity, and the concentration of wealth. They waxed indignant over notorious abuses by railroads, corporations, and political machines. Back of these instances lay the general complaint that American government, city, state, and federal, had ceased to belong to the people and had become a captive of the "interests." Where corruption had not taken hold, inefficiency reigned. The old forms of government seemed no match for new problems and new temptations. "The old political formulas do not fit the present problems; they read now like documents taken out of a forgotten age," said Wilson. To anxious citizens, Progressivism offered itself as nothing less than the urgent attempt to save American democracy in the only way it could be saved: by purifying, deepening, and updating it. In the words of the Progressive Party's 1912 platform, the movement strove "to build a new and nobler commonwealth." Reconstruction of the South may have failed, but reconstruction of the whole country and its government was now in order. As a first step, Progressives had to expose, to borrow the party platform's language again, the "invisible government" behind the "ostensible government." Behind the public forms and formalities of democracy—the Constitution and laws—reigned a "plutocracy," to use the popular term of the day, a wealthy ruling class hiding behind closed boardroom doors and mansions with high hedges. "The government," Wilson said, "which was designed for the people, has got into the hands of bosses and their employers, the special interests. An invisible empire has been set up above the forms of democracy." As a result, "The life of the nation . . . does not centre now upon questions of governmental structure or of the distribution of governmental powers. It centres upon questions of the very structure and operation of society itself, of which government is only the instrument." On the campaign trail that year he put the indictment sharply: "The masters of the government of the United States are the combined capitalists and manufacturers of the United States." The gravest threat to American democracy came from within, from the interests (not James Madison's "factions," unjust combinations that could arise either from interests or passions) who hid behind the rights of private property in order to exploit and rule the people. Imperiled by its own civil society, democracy faced an unprecedented challenge. "To destroy this invisible government, to dissolve the unholy alliance between corrupt business and corrupt politics," declared the Progressive platform, "is the first task of the statesmanship of the day."[12]

Similar indictments had been lodged by the Populist movement in the early 1890s. Despite their common hatred of "the interests" and affection for direct democracy, the two movements disagreed profoundly on the cause and cure for America's problems. The People's Party, as it was called, had diagnosed the ill as "corruption," meaning a falling away from the wise and honest standards set by George Washington and the nation's founders. "Corruption dominates the ballot-box, the Legislatures, the Congress, and touches even the ermine of the bench," as the preamble to the party's platform, adopted in Omaha, Nebraska, at its first national convention in 1892, declared. "The fruits of the toil of millions are boldly stolen to build up colossal fortunes for a few . . . and the possessors of these, in turn, despise the Republic and endanger liberty. From the same prolific womb of governmental injustice we breed the two great classes—tramps and millionaires." In a language of agrarian protest that would have been familiar to Andrew Jackson and Thomas Jefferson, the Populists denounced as "usurers" the bankers and bondholders who kept interest rates high to feather their own nests and impoverish the "plain people," that is, the "producing class," as opposed to the coupon clippers with their "gold-bearing bonds." Nor were these scheming oligarchs confined to the United States. "A vast conspiracy against mankind has been organized on two continents," the platform asserted, "and it is rapidly taking possession of the world." In writings under his own name, the preamble's author, Ignatius Donnelly, an editor, novelist, Minnesota politician, and ardent believer that Francis Bacon wrote Shakespeare's plays (Donnelly wrote two books to prove it), pointed to the "Israelites," that is, the Jews, as key actors in this international conspiracy. "The world is today Semitized," as a character in *Caesar's Column*, Donnelly's hugely popular 1890 utopian novel, explained. Officially, the People's Party kept silent on the question.[13] But in the decisive respect the Populists were reiterating or at any rate reformulating time-honored complaints about the pernicious effects of the "money power" on the "equal rights and equal privileges" of citizens. They breathed not a word of a new age that had allegedly anachronized the morality of the good old Declaration of Independence and Constitution.

What the Populists offered as a solution was the traditional resort of farmers who had gone too far into debt, or who, to put it more charitably, had the misfortune of growing crops in the great post–Civil War deflation, when commodity prices fell for decades.

The remedy was inflation, a huge increase of the money supply to be fueled by bimetallism, the "free and unlimited coinage of silver" at a 16:1 ratio to gold. In addition to this panacea, however, they proposed new steps: public ownership of the railroads, telegraphs, and telephones; a graduated income tax; the Australian or secret ballot; the initiative and referendum; direct election of senators; and various reforms of labor law, including "the further restriction of undesirable immigration," especially low-wage Chinese workers.

The future Progressives had zero sympathy with the main item on the Populist reform agenda, dismissing soft-money policy as economic moonshine, a combination of "economic disaster" and "financial dishonor," as TR once put it.[14] With the secondary prescriptions there was considerably more sympathy, and the Progressives borrowed unapologetically from the list, as did reformist Republicans and Democrats in the following decades. In journalist William Allen White's words, the Progressives "caught the Populists in swimming and stole all of their clothing except the frayed underdrawers of free silver."[15] On these smaller or less central questions the People's Party often sounded downright Progressive. "We believe that the powers of government—in other words, of the people—should be expanded (as in the case of the postal service) as rapidly and as far as the good sense of an intelligent people and the teachings of experience shall justify," the platform announced, "to the end that oppression, injustice, and poverty shall eventually cease in the land."

Yet even when advocating the expansion of governmental powers, the Populists revealed the line that separated them from their successors and later liberals. The People's Party called for expanding government's *powers*, not its majesty or superintendence over the people itself. New powers were needed to restore government to its old authority, but that authority still reported to the people and the Populists never renounced the people's prudent jealousy of governmental power. "We assert our purposes to be identical with the purposes of the National Constitution," said the Omaha Platform's preamble; even when calling for public owner-ship of the railroads, it insisted on an accompanying constitutional amendment that would put all government workers under a strict civil service system, "so as to prevent the increase of the power of the national administration by the use of such additional govern-ment employees." The national executive and central administrative power could not be suffered to expand, in short, lest they upset the Constitution's internal checks and balances and thus the people's control over their own national government. Somewhat contradic-torily, the party called for a larger administrative power without an administrative state. Though they disliked bankers, the Populists distrusted ambitious and profligate politicians as well. The coun-try's money "should be kept as much as possible in the hands of the people," safe from bankers and government officials alike, even if the latter spouted pleasant-sounding redistributionist or protec-tionist nostrums, of the sort then associated with tariff policy. The postal service reform alluded to earlier involved the creation of postal savings banks—an easy and safe way for small earners to save money at the nearest post office branch, rather than having to trust a bank. Suspicious of elites, the Populists sympathized with the workingman. "Wealth belongs to him who creates it," they de-clared, "and every dollar taken from industry without an equiva-lent is robbery. 'If any will not work, neither shall he eat.' "[16] This was not exactly the ethic of the future entitlement state. Individual Populists were sometimes crankier or more radical, but the move-ment and especially the party always came back to the principle that the powers of government should only be expanded in keeping with "the good sense of an intelligent people" and "the teachings of experience."

However much it may have paid "lip-loyalty" (TR's word) to the same standards, Progressivism in fact respected them only to the extent they jibed with a third authority, quite distinct from either popular prudence or the maxims of experience. Democratic good sense could not take the place of expertise, objected the Progressives, and mere experience could never discover the deep, systemic causes of the present discontents. What the Populists lacked above all was a *theory*, a doctrine, by which to explain the chronic inadequacy of the present distribution of government powers, and by which to prescribe the far-reaching extensions of those powers to meet the needs of the new age. Backward-looking when they should have been forward-looking, nostalgic when they should have been scientific, they could never grasp the extent to which the U.S. Constitution and its spirit were the source of late-nineteenth-century America's problems.[17] They continued to insist that the Constitution had been betrayed, when in most respects, retorted the Progressives, it had simply become outmoded. Or worse: Charles Beard, J. Allen Smith, and other Progressive historians argued that the rapacious rule of corporations, trusts, and millionaires, was more or less what the Constitution had set out to enshrine.[18] Looking back at William Jennings Bryan and the Populist movement's other tribunes, Woodrow Wilson found it hard to hide his contempt for their "crude and ignorant minds."[19]

In politics, names matter; in this case, they speak volumes. The Populists were, well, populist; the People's Party believed in the people's rule, enhanced in many respects but still broadly faithful to the constitutional ideas of the great farmer-presidents like Washington, Jefferson, and Jackson. By contrast, the Progressives believed first and foremost in a doctrine of progress. Wilson issued this panegyric to it in *The New Freedom*, still the most important book of campaign speeches in the history of modern liberalism:

> Progress! Did you ever reflect that that word is almost
> a new one? No word comes more often or more naturally
> to the lips of modern man, as if the thing it stands for were
> almost synonymous with life itself, and yet men through

many thousand years never talked or thought of progress. They thought in the other direction. Their stories of heroisms and glory were tales of the past. The ancestor wore the heavier armor and carried the larger spear. "There were giants in those days." Now all that has altered. We think of the future, not the past, as the more glorious time in comparison with which the present is nothing. Progress, development—those are modern words.

From the country's beginning, out of necessity, choice, habit, and fetish, Americans had striven to improve their material and moral condition. "The American people are not naturally stand-patters," Wilson observed. "Progress is the word that charms their ears and stirs their hearts." *Progress* was never a dirty word then, though its exact meaning was disputed and the relation between material and moral progress frequently regarded as problematic. "From the conclusion of this war," Jefferson warned in the early 1780s, "we shall be going down hill." The people "will forget themselves, but in the sole faculty of making money, and will never think of uniting to effect a due respect for their rights." Lincoln said that the Declaration of Independence contemplated a "progressive improvement in the condition of all men everywhere," but he also spoke of our "progress in degeneracy," as indifference to slavery's evil—and even proslavery arguments—gained traction before the Civil War.[20] This separation and potential conflict between moral and material progress marked a significant exception to Americans' modernity, in Wilson's sense of the word. Still, in general they regarded it as a blessing that in the New World material and moral progress were possible, and they looked forward to a future better than the present in many respects, if not all. Two additional exceptions to this "modern" outlook were, however, conspicuous. One was the old Christian insistence on the world's fallen condition, which no amount of prosperity or moral improvement could redeem. Whether worldly progress was therefore a snare and delusion, a matter of indifference, a genuine if second-rate good, a gracious sign of God's favor, or some combination, was hotly debated among and within the dizzying variety of Christian sects. To saved souls a perfectly happy future was indeed reserved, on that all believers agreed; but the blessed future was not of this world and its advent and character remained inscrutable to man here below. Such views may have been losing ground in the higher seminaries, but they remained potent.[21]

The other exception touched directly on Wilson's political aims. For Americans have from earliest times regarded the Founders and their handiwork with a certain reverence and awe. Most Americans thought there *were* giants in those days, and ever since looked up, and back, to Washington, Jefferson, and the other Founders both as venerable guides to the Republic's principles and as timeless exemplars of political wisdom more generally. As the distance from the Founders grew, so did their reputation, an effect predicted and counted on by the most far-sighted Framers, who regarded such "prejudice" as indispensable to political stability and good government. A rational constitution had to know the limits of popular reason.[22] To this "blind worship" of the Constitution and its principles, Wilson intended to help put an end. In this respect, which he regarded as the crucial one, Americans (not just Populists) remained mired in a premodern world, unable to address the very different problems of the present and incapable of imagining the "more glorious time" to come, "in comparison with which the present is nothing" and the past less than nothing. Nonetheless, the rise of Progressive sentiments presaged a new era of politics in which the blinders finally were coming off.

> We are the first Americans to hear our own countrymen ask whether the Constitution is still adapted to serve the purposes for which it was intended; the first to entertain

any serious doubts about the superiority of our own institutions as compared with the systems of Europe; the first to think of remodeling the administrative machinery of the federal government, and of forcing new forms of responsibility upon Congress.

This statement, from the opening pages of *Congressional Government*, seems to imply that the "purposes for which [the Constitution] was intended" remain a relevant standard in judging it today. By the time the reader reaches the book's final paragraph, however, the futility of such an appeal has become clear. The point is to transform the constitutional system, albeit gradually and without violence.

The charm of our constitutional ideal has now been long enough wound up to enable sober men who do not believe in political witchcraft to judge what it has accomplished, and is likely still to accomplish, without further winding. The Constitution is not honored by blind worship. The more open-eyed we become, as a nation, to its defects, and the prompter we grow in applying with the unhesitating courage of conviction all thoroughly-tested or well-considered expedients necessary to make self-government among us a straightforward thing of simple method, single, unstinted power, and clear responsibility, the nearer will we approach to the sound sense and practical genius of the great and honorable statesmen of 1787. And the first step towards emancipation from the timidity and false pride which have led us to seek to thrive despite the defects of our national system rather than seem to deny its perfection is a fearless criticism of that system.

Wilson would devote his life to that "fearless criticism," and to refounding American government as "a straightforward thing of simple method, single, unstinted power, and clear responsibility." His way of honoring the Founders was not to live up to their principles but to imitate their example by a fresh stroke of "political genius" surpassing, and superseding, their own.[23]

By opposing the special interests and ushering in new forms of government, the Progressive movement understood itself to be advancing the public interest. One of its most attractive qualities (and one of liberalism's, too) was its insistence on asserting the dignity of the public realm as over against the private world of interests. Teddy Roosevelt, Wilson, Taft, Henry Cabot Lodge, Elihu Root, and many other spirited men in that generation rebelled against the idea of a republic enthralled by the likes of Jay Gould, J. P. Morgan, and Commodore Vanderbilt. Though they despised the political influence wielded by the captains of industry, they disliked even more the popular fascination with such men, which smacked of Mammon-worship. To exaggerate slightly, there was something Aristotelian about the Progressives' indignant assertion of the supremacy of public over private good, of politics over economics, of statesmanship over the moneymaking arts. On occasion, they claimed the pedigree explicitly. Richard T. Ely, one of Wilson's principal teachers at Johns Hopkins, wrote that modern critics who demanded that economics bow to ethics had basically "gone back to the Greeks, notably to Plato and Aristotle, who subordinated all economic inquiries to ethical considerations." These "older and sounder conceptions" had been prevalent, he explained, "until a wave of revolutionary materialism in the last century swept over the world." Wilson himself, in his textbook, *The State*, stressed that "Aristotle was simply stating a fact when he said, 'Man is by nature a political animal.' "[24]

This impulse to rescue politics from its economic servitude stimulated an outpouring of new biographies of American and European statesmen—including the thirty-nine volumes of the American Statesmen Series, written by Henry Cabot Lodge, TR, William

Graham Sumner, and other leading men of the day, and published by Houghton Mifflin—not to mention an influx of talented young men (and a small but growing number of women) into politics. Compared to the politicians of our time, the leading public figures of the Progressive generation were giants—literate, deeply learned, and serious about their calling. Compared to the generation that had preceded them, epitomized by the post–Civil War presidents whose names are hard to remember and whose accomplishments are just as elusive, the Progressives appeared a return to the Republic's golden or at least silver age. For their own part, they fixed their gaze on Lincoln. Against his towering example, the politicians of the Gilded Age looked like pipsqueaks, as they would have been the first to admit. The prestige of the martyred president cast its shadow over the whole landscape of American politics. To the Progressives, who were after all mostly Republicans, Lincoln was the very beau ideal of a statesman. When he took the oath of office at his second presidential inauguration, TR wore a ring, given him by his friend and Lincoln's former secretary, John Hay, which contained a lock of Lincoln's hair clipped from his head after the assassination. But saving the Union and freeing the slaves was a hard act to follow. Leaving aside the now abandoned experiment of enforcing upon the South the terms of the Fourteenth and Fifteenth Amendments, what was left for the Lincoln-loving statesmen of the Progressive movement to do?[25]

Quite a bit, it turned out, though their efforts were guided not by Lincoln's principles or his own understanding of his career, but by the *effects* of his policies on the subsequent development of American nationhood.[26] TR regarded Lincoln as a thoroughgoing democrat and nationalist, not unlike himself. Wilson saw him as a nation-builder and frontiersman whose eye was on the future. They agreed with Herbert Croly that the sixteenth president stood for a new nationalism based on a deeper sense of democracy, now understood as the perfection of human brotherhood and therefore of man himself. The Progressives thought of themselves as defending nationalism against sectionalism, righteousness (one of their favorite words, though not one of his) against selfishness, even as Lincoln in his day had defended the Union against "the special interests of cotton and slavery," to quote Roosevelt's New Nationalism address. They sought not merely to emulate Lincoln, however, but to incorporate and transcend his work. He had been a democratic Bismarck, willing to wage total war in a "supreme and final struggle" between the "forces of health, of union and amalgamation" and the "forces of disintegration." As a result, the United States emerged as a real nation, a united or unitary political community, for the first time. Alas, its constitution and political principles, its "literary theory," remained frozen in the past. American democracy remained largely a formal thing, often a mere matter of votes and the dollars to buy them. Political democracy needed to be extended and deepened; it fell short of social democracy and social justice; its moral tone wanted righteousness. This "new nation," in Wilson's phrase, lacked a proper State to guide, but also to express, its spirit. The great Progressive task thus involved replacing the country's obsolete, eighteenth-century government with a modern State, bringing the United States to completion, finally, as a mature nation-state.[27]

## The Fourth Branch of Government

Building that State was the ultimate purpose of Wilson's "new political sentiment and party." It was no coincidence that he came to this resolve while an undergraduate at Princeton and restated it after graduate study at Johns Hopkins. For the "defects" of the Constitution and the virtues of the State that might replace it were persistent themes of his university education. Before turning to the character and teachings of this new science of the State, we ought to consider the organs that transmitted it, namely, the research universities that began to appear in America after the Civil War. The rise of the university changed not only American higher education but also American politics, in ways that are not widely recognized. Run your finger down the front page of the daily newspaper or down the column of a news Web page, however, and you will see evidence of the changes. In very many stories you will see reference to one or more "experts" offering up an expert opinion on some question or other—the causes of high unemployment, cures for juvenile delinquency, the future of terrorism, etc. Most of the experts will be professors, or Ph.D.s who work in think tanks or in government, or if not Ph.D.s then holders of comparable degrees in the professional schools of law, business, or medicine. Before the rise of the university, America was a country almost entirely bereft of such experts and of such expertise. It was a rather different America.

Not that the United States lacked higher education, for even at the time of the Revolution the Americans boasted more colleges than their erstwhile mother country. Prior to the last third of the nineteenth century, however, these schools overwhelmingly were liberal arts colleges of one sort or another, often connected to churches, and "in some respects," note Christopher Jencks and David Riesman, "more like today's secondary schools than today's universities." They did not have faculties of specialized scholars organized into departments; "everyone taught almost everything"; few teachers had had any sort of advanced education beyond a B.A., because such education wasn't available in America, except to a small extent in theology and law. Consequently, the notion of "a college professor as an independent expert with a mission transcending the college where he happened to teach was almost unknown." Until late in the nineteenth century, colleges probably had much less influence, cultural and political, than churches did, though in the formation of individual minds—for example, Madison studying with John Witherspoon at the College of New Jersey—a college education could be a significant influence. But then Witherspoon was also a clergyman, an exception proving the rule. If any jobs required a B.A., they must have been few and far between; hence the economic benefits of higher education were small compared to today.[28] For the irrelevance or at least limited influence of colleges in those days the most striking symbol is the speeches at Gettysburg on November 19, 1863, at the dedication of the Soldiers' National Cemetery. Edward Everett, not wearing lightly his Harvard A.B. (he was class valedictorian) and Prussian Ph.D. (the University of Göttingen), and being not only a former Harvard professor but also a former president of Harvard College, spoke first, for more than two hours. His speech began with a long discussion of ancient Athenian funerary rites, and proceeded to a detailed exposition of the Battle of Gettysburg. The second speaker, Abraham Lincoln, who estimated "the aggregate of all his schooling did not amount to one year," spoke for about two minutes.[29] Who gave the more significant address?

With the rise of the university began a new chapter in American history. The change came slowly: the first Ph.D. was granted in 1861 by Yale, and in 1869 Charles W. Eliot began his forty transformative years as president of Harvard. When he introduced the elective system, it not only empowered students but also encouraged the faculty to organize itself along more scholarly and specialized lines. But it was not until the founding of Johns Hopkins as a research university in 1876 that the modern university, dedicated to uncovering new knowledge discipline by discipline and focused more on graduate than undergraduate education, put down deep roots. Graduate departments and schools proliferated in the 1880s, including at leading state universities like Michigan and Wisconsin, and by the 1890s, with the founding of the University of Chicago and reforms at Columbia and Princeton, the research university had established itself. By Wilson's presidency the two dozen or so major research schools, still recognizable as such today, were in place.[30]

Between the Civil War and World War I, the reformers of the American academy looked to Germany, especially to Prussia, for inspiration. There were exceptions—Cornell's founders, for example—who took Oxford and Cambridge as principal models. But Germany had the most advanced, well funded, and comprehensive institutions of higher education in the world, and Americans had a penchant for studying there that stretched back to the early nineteenth century, when a handful of Americans including Everett and the historian George Bancroft ventured to Germany for advanced degrees and a de rigueur visit with Goethe in Weimar. In addition, Germans emigrated to the United States and began teaching here, notably Francis Lieber, who arrived in 1827 to begin his career as an author and educator at South Carolina College and later Columbia, where he became America's first professor of political science. After the Civil War, these rare birds were joined by a much larger migration of young Americans who went to Germany to study (access was easy and inexpensive, like the beer) and came home sporting German Ph.D.s. They returned as evangelists for the German system of higher education, too, and helped encourage and staff America's emerging research universities. A who's who of Progressive intellectuals, this cohort educated, in turn, a second generation of thinkers, who didn't need to escape to Germany to earn graduate degrees. Wilson was a member of that second generation, trained at Johns Hopkins by German-educated scholars, particularly the historian Herbert Baxter Adams and Ely, an economist-cum-sociologist who was also a pioneer of the Social Gospel movement.[31]

What above all attracted the Progressives to the German academy, however, was its direct connection to political life. Before the Left's avant-garde became captivated by the Soviet Union as the model society of the future, it fell in love with Germany. In Germany all universities were owned and operated by the State, and all professors were civil servants, many of whom, especially the professors of political science, served as advisors to the government and as legislators. This marriage of knowledge to power proved immensely attractive to the young men who had gone abroad seeking both. Many were struck by the planned efficiency of Prussia's wars to unify the German nation—as against the bloody, protracted mess of the American Civil War.

It's amusing to see their puppy love in action. Frederic C. Howe, a Johns Hopkins Ph.D., wrote a book celebrating Robert M. La Follette's Wisconsin as "an experiment in democracy," marked above all by the close connection between the state government and the state university. "Social science in many of its basic dimensions began as a reform movement," the political theorist John Gunnell observes, and nowhere was this truer than in Wisconsin. Howe began his paean: "Wisconsin is doing for America what Germany is doing for the world."

> Just as the German burgomaster builds with a far-seeing vision to promote the comfort, the convenience, the health, the beauty of the city; just as the German empire has been consciously developing the education, training, and efficiency of its people, just as the state-owned railroads, waterways, harbors, forests, and mineral resources are used for the upbuilding of the fatherland . . . so Wisconsin is building a commonwealth in a conscious, far-seeing, intelligent way. It is becoming an experiment station for America.

In this epic of Germanification, the University of Wisconsin at Madison had a crucial role.

> The state university . . . is a scientific research bureau, using its faculty and equipment in the service of the state. Professors are connected with almost every department of public administration. State problems are studied in the schools of politics, of agriculture, of mechanical

engineering. Experts from the university are employed on
railway, taxation, and industrial problems, and in extend-
ing the influence of the university throughout the state.
. . . The university is largely responsible for the progressive
legislation that has made Wisconsin so widely known as a
pioneer.

And what of popular government?

Scientific thoroughness characterizes politics as in no
other place in America. Legislation is preceded by exact
knowledge of the abuses to be corrected and the ends to be
achieved. Laws are made as simple and direct as possible.
The politician has almost disappeared from the state-house.
He does not thrive in this atmosphere. There is little partisan thinking, and little partisan legislation.[32]

Although "calling in the expert," as Howe phrased it, accomplished some good in Wisconsin, it would be naïve not to acknowledge the threat to equality and the consent of the governed posed
by self-proclaimed experts in public administration whose new-
found power the Progressives found so exhilarating. "Wisconsin
is making the German idea her own," Howe wrote. "The university is the fourth department of the state. . . . There is no provision for this in the constitution, no reference to it in the laws."[33]
Exactly. A distinctive mark of the Progressive movement was the
aspiration to turn the university into the fourth branch of government, with no constitutional or legal limits on its informal powers.
That Prussia was a kingdom with an entrenched aristocracy, that
Germany was a not-very-constitutional monarchy, seemed not to
discomfit those who wanted to import Teutonic institutions. Wisconsin wasn't Prussia, to be sure, though a lot of Germans were
loose in each. But the presumption was that the university could

be trusted to define the state's social problems, prescribe cures, and then evaluate the success of the very programs it had recommended. Everything depended on the integrity and wisdom of the experts, which could only be judged by . . . the experts, with perhaps a few of the vanishing politicians looking on weakly. Tenure and academic freedom (*Lehrfreiheit*, the Germans called it) tended to reinforce the argument's circularity. Political pressure on universities and faculty members wasn't uncommon in the nineteenth century, but it decreased as the university ideal took hold.[34] Still, nothing separated Progressivism from Populism, or for that matter from all previous American democracy, more sharply than this faith in the presumptive expertise, integrity, and political authority of the academic mandarins. Though the civil service is often called the fourth branch of government, the bureaucrats form, in a larger perspective, a subdivision or extension of the academy reflecting experts' viewpoints and interests, more faithfully than the public's. For all of its emphasis on the people versus "the interests," the Progressive movement did not trust the people to govern themselves unless under the influence of the "exact knowledge" of an expert class.

## The Doctrine of Progress

The modern State and the modern university thus had a symbiotic relationship from the beginning. They were two parts of the same reform project. High on the agenda of the builders of America's research universities, therefore, stood the creation of academic departments to house the new social sciences, which were essential to the university's (and the State's) task of experimenting on society, that is to say, on the people. A great deal of effort went into building up the natural sciences as well, and soon Ph.D.s in chemistry, engineering, agronomy, and other fields poured out of the graduate schools. In a democracy that had always been fascinated by inventions and improvements and the pursuit of material well-being, these academic developments made perfect sense, and for the most part needed no new theory to justify them. Yet the new social sciences—economics, psychology, sociology, history, anthropology, and political science—based on new assumptions about man and society, formed a different case. Far more unified then than now, they stood at the center of Progressive reforms. In these departments especially, the purpose of the university as a whole, which meant its place in modern society, came into sharp focus.

Wilson did not discover, much less invent, these new social sciences. He studied them in college and graduate school, and then modified them until he had shaped his own version of political science—which by precept and example, as we shall see, he taught to modern liberalism. Broadly speaking, these disciplines all conceived of themselves as historical sciences, in a sense that none of their precursors would have recognized. Whether discussing economics, family structure, race, religion, or forms of government, they assumed that the starting point of science was "the historical sense," the notion that all thought is a child of its time. As Wilson wrote nonchalantly, "The philosophy of any time is, as Hegel says, 'nothing but the spirit of that time expressed in abstract thought'; and political philosophy, like philosophy of every other kind, has only held up the mirror to contemporary affairs."[35] History is not merely a record of what has happened (one damn thing after another, said a historian famously), but is something hitherto unknown, a dimension of reality shaping or determining all thought and thoughtful action in an epoch, like an invisible magnetic force moving iron filings. Even as the power of magnetism had to be discovered, so did the force of history.

Whereas a magnet arranges filings in the same pattern, however, history marks every epoch as unique. Yet these differences link up, because history isn't the realm of chance, fate, and accident, but of order, specifically of a kind of orderly development. History with a capital $H$ shows that out of the apparent disorder of human events, order emerges, without being intended by anyone. Connecting age to age, history's logic arranges all events and cultures into a rational sequence from the simplest to the most complex, from the least free society to the most free. As society changes, so does the human mind and human nature with it. In fact, the mind changes first, though it isn't fully conscious of the change. Over thousands of years man becomes more aware of himself, of his freedom and reason. He grows up. The process might be called teleological, except that the end was not visible or foreseeable from the beginning, as one knows an acorn will become an oak tree and nothing but an oak tree. For millennia man did not, and could not, know he was meant for freedom: the ancient Athenians saw no contradiction between slavery and their democracy, according to Hegel. Caesar could not have known that by conquering Gaul he made inevitable Rome's conquest centuries later by Germanic tribes, thus ushering in the Christianization of Europe, feudalism, the Renaissance, the nation-state, the Protestant Reformation, the Enlightenment, the French Revolution, and the final emergence of universal freedom in the rational State—but inexorably each event led to the next.[36] Though these developments couldn't be known in advance, and each was the occasion of passionate and bloody conflict, they make perfect sense in retrospect. Hence it's not man's *nature* but his *history* that explains what he is today.

G. W. F. Hegel (1770–1831) was the philosopher who first articulated this notion of history as the story of man's necessary development, or progress. In many ways he laid the deepest underpinnings of modern American liberalism, though his thought had to be adulterated and democratized before that could happen. Conservatives often mistake Hegel for a relativist because he argued that morality, like all thought, changes with the times. But the changes ceased with the end of history, when philosophy culminated in wisdom and true morality. Hegel's so-called relativism ended in absolutism, perhaps the severest and most self-satisfied in the history of philosophy.

His lectures on the philosophy of history take the reader on a Cook's tour of world civilization, traveling from east to west (after a short stop in prehistoric Africa), with visits to the "oriental mind," the Greek and Roman minds, and finally the Germanic one. For history to be rational it had to be a work of mind; but mind could be historical only if it were a collective mind or spirit (*Geist*) that could be embodied in a historical age. Hegel's famous concept of the *Volksgeist* met these needs. Each culture or folk mind (e.g., classical Greece) passed from youth to maturity to decline, but handed the torch to a successor just in time. The essence of the fading culture was preserved in the transfer, so that nothing important was lost and the subsequent folk mind could begin where the last left off. And so the world mind (*Weltgeist*) progressed, from the consciousness, as Hegel put it, that One is free (the principle of oriental despotism), to the understanding that Some are free (the Greeks and Romans), to the knowledge, finally, that All human beings are free (the Christian insight, perfected in the Germanic mind). This change in moralities, according to Hegel, was the most highly moral thing in the world.[37]

The debt that Progressive idealism owed to German idealism was enormous. Progress in the Hegelian sense bred the lofty assurance that history had a meaning and direction, that it pointed inevitably to human liberty and human flourishing. Small wonder that many Progressives found it so easy to identify change with improvement. If history guaranteed an ethical and intellectual consummation, after all, how dangerous could change or experiment be? Almost any change had the character of improvement, they figured, or at least greased the skids for improvement. The same idealistic reasoning undermined or at least dangerously weakened the case for political moderation: why settle for less now, when soon, very soon, you can have it all? Why tolerate imperfect justice when perfection is within reach? The feeling of righteousness that came from being on the right side of history, and being applauded for being on the right side of history, could easily substitute for the deep self-knowledge that had always been the accompaniment and requirement of true righteousness.[38]

These temptations arose partly out of Hegel's own system and partly out of modifications of it by his students and their epigones, including the Progressives. After all, Hegel taught that progress came to an end. This hard doctrine was compelled by his premise that all thought is a child of its time—which meant that Hegel's thought was a child of *his* time, no less relative to the age than any previous thinker's. Why then was his notion of the rational State any less preposterous or mistaken than the divine right of kings, or suttee in India? The difference was not in the principle of historical relativity but in the nature of the times. This was the significance of his elusive idea of the "end of history," revived and made fashionable by Francis Fukuyama after the end of the Cold War. Hegel's teachings could be true because he philosophized, at the "Absolute moment" when relative knowledge ceased and absolute knowledge began. (It was still not easy, of course, to become the first Wise Man, but Hegel was up to it.) Truth remained relative to history, but when history turned Absolute, so to speak, so did truth. Nonetheless, truth would be off to the races again if the conditions of Absolute knowledge did not persist, if history as the story of man gaining maturity and wisdom did not have a Finis at the end. Hence the *end of history*, which didn't mean that births, deaths, and other events ceased, or that scientific inventions and artistic productions stopped. It meant that no important ethical, philosophical, and even political truths remained to be discovered or understood anymore.[39]

The flip side of the end of history was the advent of the "rational State." Hegel saw it, in outline at least, in early nineteenth-century Prussia. History culminated in this final and highest form of human government, which protected the freedom of every individual equally but was a constitutional monarchy, not a democracy.

To be rational, you see, government had to be confided to "those who know," not to the many who don't. Most countries were not ready for this consitution, but in due course all would be.

It would be going too far to say that what remained to be accomplished was merely administrative, but it was close enough for a certain dispiritedness to set in among many of Hegel's followers. The end of history meant the end of philosophy or of the quest for wisdom, and the beginning of the reign of complete wisdom, world without end, amen. This implied the end of thinking, in the Socratic sense; and the end of obedience, in the biblical sense, to a mysterious God who was separate and apart from the world and who existed in eternity, not time. The wisdom now available to man was entirely of this world and free, absolutely free, of all skepticism and humility; it could be deemed superhuman, except that it came into existence by or through man's own unconscious labors over the ages. Its possession depended on the completion of a historical process over which no thinker or philosopher had control, and by which his or her own "thought" decisively had been controlled. To that extent Hegel's wisdom was at once superhuman (like a revelation or a miracle) and subhuman (like a deterministic response to an outside stimulus). What it wasn't was properly human, suitable for fallible and free men who are neither gods nor beasts. Once government could be based on such wisdom, it no longer needed to be *limited* due to man's intellectual and moral imperfections.[40]

So the end of philosophy also meant the end of politics and liberty, in the sense of deliberating over opposing principles of justice and the public good. Those grand disputes belonged to the past, when man could only desire to know; now he knew, and could see with his own eyes the clashing theses tamed into syntheses. Nevertheless, wars between nations and poverty within nations would continue to occupy public attention, Hegel thought, and his students made the most of it. Still, how interesting could such residual issues be when history itself was over? Alexandre Kojève, Hegel's greatest twentieth-century interpreter, put his own existentialist-Soviet spin on the dispiritedness at history's end: "There is no longer fight nor work; History is completed; there is nothing more to *do*. . . ."[41] Politics was doomed to be replaced by administration, by bureaucratic management of the society and economy. Hegel's followers, the most impressive ones at least, rebelled against this life sentence without parole. His acolytes on the Right doubted that the Germanic folk mind—the modern West—could long persist in such an unhealthy state. None of its historical predecessors had. Like them, the West seemed doomed to decline, rationalized Christianity to decay—though hugely different accounts of that decline issued from Nietzsche, Max Weber, and Oswald Spengler.

On the Left, Karl Marx tried to save progress, temporarily, and even a form of idealistic striving, from history's pitiless end. Unable to reject the Hegelian framework, he settled for upending it. But turning Hegel on his head meant putting his feet on the ground, inasmuch as the master had tried to build his system on ideas, on reason, on intellectual production rather than material production. By putting economic change in the driver's seat and relegating *Geist* to an effect or by-product of the mode of production and its resulting class relations, Marx invented the most powerful form of left-wing economic determinism. And he helped confuse American Progressives ever afterward over whether socioeconomic development drove politics, or vice versa. Most important, he revised the understanding of the so-called end of history. The end is near, Marx argued, and it's close enough that the more sensitive among the proletariat and its vanguard may discern some of the arrangements concerning the initial transition from capitalism to socialism. But it's not close enough to understand, except in the vaguest way, what the actual end of history, the stateless society of communism, will be like. That heaven remains beyond our ken, but not—and this is crucial—beyond our striving. By locating man on the suburbs, perhaps even the distant exurbs, of paradise, Marx created space for one last great political action, one final rampage of ruthless statism, one sweetly dying gasp of idealism, before politics and idealism were retired to the museum of proletarian oppression.

The Progressives were not great admirers of Marx and in fact abhorred his strategic recommendations—fanatic partisanship, class warfare, and violent revolution—as hopelessly reactionary. Yet like the Marxists, the Progressives needed an account of history that would explain progress—the rationality of the historical process—without ruling out future installments. So they borrowed from the Marxists the notion that the moment of truth, the moment when Truth emerged from ideology and when the relativity of all previous thought became clear, could be separated from the full-fledged reign of wisdom at the end of history. The truth about past thought and action, argued the Progressives, could be well enough understood in light of the present, even though the present as yet caught only a glimpse of the future. Having wrestled with his own passions, Augustine would have recognized the Progressives' dilemma: bring History to an end, O Lord, but not yet. More honest or at least more impatient, the Marxists enlisted Revolution to close the gap once and for all between history and the end of history, between the world as it is and the world as it should be—and soon will be. The Progressives, however, replaced Revolution with Evolution, and set out not to advance or force the end of history but to endlessly approximate it. Their approach to the perfect society would necessarily be asymptotic, as the mathematicians say: the curve of social progress would touch the line of complete social justice only at infinity. The scheme had the great advantage of giving Progressives something to do *forever*. And it kept the old idealism up: from any point on the curve, no matter how close to perfect social democracy, infinite progress was still possible, and yet more obligatory.

As attractive as this prospect was, it evaded rather than solved the problem that a morality claiming the high ground of history had finally to command that ground. Progress, as opposed to mere change, must have a fixed point or standard by which to be measured. If the goalposts are constantly receding, you aren't making progress. (Wilson liked to quote *Through the Looking Glass* as "a parable of progress." Out of breath after a furious race, Alice exclaimed, " 'Why, we are just where we were when we started!' 'Oh, yes,' says the Red Queen, 'you have to run twice as fast as that to get anywhere else. . . .'")[42] If History never ends, then it cannot be shown to be rational, much less happy or moral. Unending progress is what Hegel condemned as "bad infinity," no progress at all. The Progressives took this problem and turned it into an opportunity— for endless State-building. They called to their assistance the other major intellectual current of the era, Social Darwinism.

## The Evolution Revolution

Charles Darwin published *On the Origin of Species* in 1859, the year after the Lincoln-Douglas debates. By the time of Lincoln's presidential inauguration the book was being taught at Harvard, and Asa Gray, its leading proponent on the faculty, was explaining it in a series of articles in the *Atlantic*. In a way, Darwinism was old news. Charles Lyell and other geologists, studying rock sedimentation, had proposed in the 1830s that the earth was a product of slow evolutionary forces, and hence older than the orthodox Christian dating of the world. Even earlier, in his *Second Discourse*, Rousseau had sketched a speculative history of a process, entirely accidental, by which man over many eons had emerged from non-man. Darwin didn't renounce speculation entirely but he emphasized the data points in the fossil records and among living species, connecting the data by an account of evolutionary change as the law of life. Despite its emphasis on slow, marginal alteration, Darwin's theory had a revolutionary effect; it hit hard and fast, an Evolution Revolution. As John Dewey said fifty years later, Darwin accelerated "the transfer of interest from the permanent to the changing." Darwinism did not merely question the truth of the old notion of unchanging species; as a cultural phenomenon, it threatened to make obsolete the whole search for unchanging truths. *On the Origin of Species* "conquered the phenomena of life for the principle of transition," explained Dewey, and its effect on biology inspired the transformation of philosophy and the social sciences, too. It "freed the new logic for application to mind and morals and life." Darwinism did not refute the older natural philosophy, much less the biblical view of the world. It rendered them passé. Then and now, natural selection had Christian adherents who argued that its evolutionary effects were secondary causes by which the First Cause acted. But neither Darwin nor his followers were terribly interested in the possibility. As Dewey commented presciently, "intellectual progress usually occurs through sheer abandonment of questions . . . that results from their decreasing vitality and a change of urgent interest. We do not solve them: we get over them."[43]

Social Darwinism, the application of Darwinism to social problems, is usually depicted as a smug phenomenon of the Right, an elaborate rationalization of capitalism by its beneficiaries and supporters in the Gilded Age. Andrew Carnegie's name invariably comes up, and William Graham Sumner's, the iconoclastic professor of sociology at Yale. In fact, Carnegie, a follower of Herbert Spencer, was one of the few millionaires who could be considered anything like a self-confessed Social Darwinist. (It didn't prevent him from being a major philanthropist.) Most businessmen in those decades, according to historian Paul Boller, "explained themselves to the world in terms of hard work and Christian stewardship," not the survival of the fittest.[44] The standard accounts conjure and condemn Social Darwinism as a plutocratic ideology, despite the fact that Sumner and many others who defended laissez-faire economics despised "plutocracy," which they associated with the high protective tariff and what we'd today call crony capitalism. Laissez-faire economics long predated Darwin, of course, who said in fact that the idea of natural selection had come to him from reading Malthus's *Essay on Population*. This identification of capitalism and the so-called robber barons with Social Darwinism is, however, a conceit of modern liberalism: a tendentious simplification that permits historians to celebrate the Progressives as humanists and moralists who were withal principled foes of the cruel Social Darwinists. The two sides in this liberal morality play are readily identifiable even today in Barack Obama's speeches. He enjoys nothing so much as abominating the "trickle-down economics" of the "Social Darwinists" who deny the moral truth that they are their brother's keeper. "There's something bracing about the Social Darwinist idea," he allowed once, "the idea that there isn't a problem that the unfettered free market can't solve." As a philosophy it means "every man or woman for him or herself," he said on another occasion. The problem is it "requires no sacrifice on the part of those of us who have won life's lottery," implying that the Social Darwinist test is not one of fitness but of plain dumb luck.[45]

For these purposes "Social Darwinism" stands in for the sheer greediness and low glower of the American bourgeoisie. It functions as a term of opprobrium, conveying contempt for the business end of American civilization, not merely for the millionaires but for the millions on the make. One would never guess from its usage now or then that the more influential and probably the more numerous Social Darwinists were *on the Left*, not the Right, that the new social sciences in the new universities were suffused with the Darwinist spirit, and that Progressivism depended on it for an important component of its doctrine of progress.[46] At this point we need to invoke another scene from *Through the Looking Glass*—the famous exchange in which Humpty Dumpty tells Alice, "When *I* use a word, . . . it means just what I choose it to mean—neither more nor less." For to the liberal mind Social Darwinism is immoral, and the respectable Darwinism of the Progressives has got to be something else: call it "Reform Darwinism," or evolution-ary social science, or "the method of intelligence."[47] Their squea-mishness is all the more paradoxical because modern liberals pride themselves on defending Darwinism against the yokels and yahoos on the Religious Right who allegedly want to suppress it; liberals have delighted in replaying the Scopes Trial, well, ever since the Scopes Trial. But they fought hard to keep "Social Darwinism" as a term of abuse against the defenders of free-market economics. More slyly, they used the term to suggest that the opening debate of the modern era was between antigovernment ideologues who lacked a sense of humanity, on the one hand, and warm-hearted pragmatists keen to use science to alleviate human misery, on the other. And granted, "Social Darwinists" like Sumner were exactly the kind of opponents the Progressives would choose for themselves if they could—and so they did, over and over again, in their own accounts of their history. Shoved out of the picture was the main-stream of the American political tradition, which for a century already had debated the size and purpose of government and the provision of social welfare in terms of natural rights and republican self-rule, rather than in terms of competing Darwinian imperatives.

The truth is that there were always many varieties of Social Darwinism, depending on the unit—individual, class, race, nation—thought to be waging the struggle for existence, and de-pending on whether human progress was thought continuous or discontinuous with natural selection. For instance, laissez-faire advocates envisioned the individual competing against everyone and everything, Marxists and a few Progressives looked to classes as the relevant antagonists, but most Progressives embraced either race or nation as the key (TR and Wilson embraced both). The two best-known American sociologists split over whether human evo-lution remained subordinate to natural forces—Sumner regarding Malthusian scarcity, the "land-man ratio," and human passions as inescapable constraints; and Lester Frank Ward holding that once man had discovered the principle of evolution, he could master it and henceforth control his own destiny free of natural determination. Almost any dispute between "Social Darwinists" and Progressives, therefore, was actually a dispute among different types of Social Darwinists.[48]

Out of this school of thought, broadly considered, came two signal contributions to Progressivism. First, Social Darwinism eased, by denying or simplifying, the end-of-history problem. The Absolute moment now could be translated as Darwin's discovery of "the origin of species by means of natural selection, or the preservation of favored races in the struggle for life," to use the full title of his blockbuster. Once the principle of evolution had been understood, much of previous science, religion, and philosophy crumbled into dust. The nature of nature could now be seen clearly; the age of Absolute Knowledge had begun. Even though Darwin's breakthrough came on behalf of "the principle of transition," as Dewey called it, that principle was itself not subject to transition. Like the Hegelians, the Darwinians exempted themselves from their critique of all previous knowledge. However much all knowledge or all scientific knowledge was now understood to be provisional, changeable, experimental, pragmatic, even socially constructed, this understanding of science was itself absolutely true, and final. No evolution past the principle of evolution was possible. Despite the insistence on a Darwinian Absolute moment, Social Darwinism attempted by and large to dispense with the end of history: "the struggle for life" was never ending, and its point was the straightforward one, for the organism to survive, or to survive long enough to breed. History did not need a rational peak or point of fulfillment beyond that.

Or did it? As biology and history became one, the truth of progress took on a new solidity and, indeed, loftiness. Not human experience alone, but the whole book of nature testified to the inevitability of progress. Herbert Spencer—the British sociological guru who actually invented the phrase "survival of the fittest" (Darwin regretted not incorporating it into *Origin of Species*, and added it in subsequent editions)—proposed a kind of unified field theory that made the principle of evolution the key to all nature and thus to all science. He set out in his ten-volume *Synthetic Philosophy* to integrate biology, geology, and even cosmology with ethics, psychology, sociology, economics, and history. The grand synthesizing principle was the law of progress that he adapted from a German biologist's description of the development of an egg: "Evolution is a change from an indefinite, incoherent homogeneity to a definite, coherent heterogeneity, through continuous differentiations and integrations." In other words, man and the universe both developed in stages from simple to more complex; that was the law of both nature and history. A British mathematician soon hatched his own devastating parody of Spencer's principle: "Evolution is a change from a nohowish, untalkaboutable, all-alikeness, to a somehow-ish and in-general-talk-aboutable not-all-alikeness, by continuous somethingelseifications and sticktogetherations."[49]

Nonetheless, Spencer's ideas exerted a huge influence on Carnegie, Sumner, and Wilson, among others. Even Ward, who broke with Spencer and Sumner over the "psychic factor" in evolution, shared Spencer's confidence that man was no exception to nature's ever upward movement. In *Dynamic Sociology* (his reply to Spencer's *Social Statics*) and in his own know-it-all six-volume *Glimpses of the Cosmos* (originally planned to fill a dozen volumes), Ward bent Social Darwinism in a "reform" direction, arguing that with man's evolution into a rational animal, human reason had become an evolutionary force unto itself, capable of accelerating and perfecting human nature and society rather than depending on the wasteful ways of nature. Man was capable of rational planning or "telesis," as Ward called it, and planned evolution or "artificial selection" would be incomparably more efficient and humane than natural selection, even as animal eugenics was much more successful than nature's own lax breeding habits.[50]

Though Darwinism shared Hegel's confidence in the unplanned emergence of order, most of its variants cut against his insistence that human consciousness was fundamentally different from, and higher than, animal consciousness. His account of history as the development of freedom, including freedom from natural determination, didn't comport with the usual mechanisms of natural selection; and his view of the end of history as the ethical and intellectual peak differed in kind from Darwinism's grim account of survival as the natural measure of all things. Yet Ward's emphasis on the radical emancipation of mind from nature, and Spencer's elevation of nature's purpose from mere survival to a certain kind of complex flourishing, helped smooth the way for the strange confluence of freedom and determinism, of German idealism and evolutionary science, that marked Progressive thought. Despite ample contradictions and blind spots, this thought would become the seedbed of modern liberalism.

## The Living Constitution

Woodrow Wilson wasted no time in putting these doctrines to work. As Princeton's president, invited to lecture at Columbia University in 1907, and as a candidate for the U.S. presidency five years later, he expounded a new theory of politics to diverse audiences in similar language. The only difference was that when speaking as an academic, he silently attributed the account to himself; when speaking to the people, he attributed it, or at least its organizing thought, to someone else, "a very interesting Scotsman" who visited him one day at Princeton. As a politician, it's usually better not to seem original, especially if you are introducing new modes and orders to an established country, aspiring in effect to become its refounder.[51] In any event, Wilson attributed not to Hegel but to the unnamed Scotsman, an expert on seventeenth-century philosophy, "the fact that in every generation all sorts of speculation and thinking tend to fall under the formula of the dominant thought of the age." For the American Founders, the *Zeitgeist* had formed under the influence of "the Newtonian Theory of the universe"; nowadays, "since the Darwinian Theory has reigned among us," everyone is likely to express everything "in terms of development and accommodation to environment." "In our own day," he told students at Columbia, "whenever we discuss the structure or development of anything, whether in nature or society, we consciously or unconsciously follow Mr. Darwin. . . ." The Founders' understanding, "a sort of unconscious copy of the Newtonian theory," was the best available at the time, but remained "sort of unconcious" precisely because it was the product of its times, which weren't end-times. The Founders couldn't appreciate "that government is not a machine, but a living thing," and so couldn't grasp the proper limits of Newtonianism, which was true partially, that is, for physical and mechanical systems alone. Philosophical captives of their age, the Founders understood all nature to be a realm of regular, predictable motion according to fixed laws. Nature meant matter in motion, obeying the law of gravity; in the case of human matter, obeying also the law of self-interest. With their eighteenth-century blinders, they couldn't see that matter was ultimately obedient to Spirit, and to Spirit's motion or evolution in history. The Constitution's Framers were "scientists in their way," but merely "the best way of their age." By the early twentieth century, however, men could know, and know that they knew, with certitude. Darwinism, including its view of government as a kind of organism, Wilson declared, "is not theory, but fact."[52]

True to their theory, the Founders "constructed a government as they would have constructed an orrery," one of those miniature solar systems that move when you crank them, every planet revolving in fixed relation to every other and to the sun. "The Constitution was founded on the law of gravitation," Wilson said, pointing to its "mechanically automatic balances" that held every department rigidly in place, even as the solar system is kept orderly by reciprocal gravitational forces. Congress checked the president, the president checked the Congress, the judiciary checked both of the elected branches, the state governments balanced the national government, and so forth. With separation of powers, checks and balances, and federalism, not to mention Montesquieu's discourses and *The Federalist* to explain it all, American politics turned into a Darwinian nightmare. This celestial or mechanistic theory produced a government that tended to go in circles (ellipses, for you sticklers) instead of forward. It wasn't progressive. It didn't evolve. And that, according to Wilson, was exactly what its authors had wanted.[53]

Fortunately, they didn't know what they really wanted, or needed. Being "practical men," they were bad theorists; they "made no clear analysis of the matter in their own thoughts." Montesquieu, whom Wilson regarded as the Founders' chief philosophical guide, had written a very long and complicated book, which they followed as faithfully as they could; but they had a government to found. "Though they were Whig theorists," the Framers "were also practical statesmen with an experienced eye for affairs and a quick practical sagacity"; their unconscious Newtonianism left room for a Darwinian cunning, or perhaps a Hegelian insight into the future. They bequeathed America "a thoroughly workable model," which has had "a vital and normal organic growth," due not to the model itself but to its gaps and omissions, its incompleteness, and especially to its openness to have "the influences and personalities of the men who have conducted it and made it a living reality." In short, the genius of the Framers had been to make a government of men, not of laws, a government open to change despite itself. Their theory had been "inorganic," their practice "organic," as Wilson summarized matters. If the Constitution's theory or principles had been followed, or followed more strictly, the country "would have had no history." That is, it would have died or, at best, deteriorated into a changeless, steady-state society like those in the distant past.[54]

But the problems facing the country had multiplied and intensified in the new age dominated by the special interests, mass immigration, and mass production. The workarounds that had sufficed in the previous century were increasingly inadequate to the twentieth. Whereas the conflict between the Founders' and the Progressives' political science didn't need to be and couldn't be articulated previously, it had to be insisted on now. It had to be confronted on the level of theory, even though Wilson and the Progressives sometimes pleaded that they had no "theory," strictly speaking; they had history, and thus facts or actualized concepts, on their side.

This ambiguity was part of Wilson's rhetorical strategy. He could in one breath call for revolution based on a new theory for a new age, and in the next recoil in horror at the very mention of theory, much less revolution. "Why are we in the presence, why are we at the threshold, of a revolution?" he asked on the campaign trail in 1912. "The transition we are witnessing," he cautioned, "is no equable transition of growth and normal alteration; no silent, unconscious unfolding of one age into another, its natural heir and successor." On the contrary, society was radically questioning its foundations and stood "ready to attempt nothing less than a radical reconstruction," which could be held back from becoming a revolution only by "honest counsels" and "generous cooperation." Probably no age "was ever more conscious of its task or more unanimously desirous of radical and extended changes." Yet far from auguring a "bloody revolution," these desires for radical change could usher in a "silent revolution" in which America recovers "in practice those ideals which she has always professed."

He could, and did, go on like this for pages. Advocating revolution, counseling cooperation lest the revolution turn violent, then denying that the coming political change will be radical or revolutionary at all, *except* perhaps in the sense of the last act of the American Revolution, in which the Framers "put aside" the Articles of Confederation in favor of the Constitution (in effect, a real change of regime, though not advertised as such)—he made sure to raise expectations of radical reform while covering himself against any hint of promoting radicalism. Up hill and down hill his prose scampered, promising a radical break with the past that would be anything but a radical break with the past.[55]

So after he warned us, for instance, that this will be "no equable transition of growth and normal alteration," no "silent, unconscious unfolding of one age into another," we find him affirming a page later that this will indeed be "a silent revolution," and fifteen pages later that "you cannot tear up ancient rootages" or "make a *tabula rasa* upon which to write a political program. . . . You must knit the new into the old." In a much-quoted statement, he went on to say, "If I did not believe that to be progressive was to preserve the essentials of our institutions, I for one could not be a progressive." It depended, of course, on what one considered essential. His carefully calculated alternations have bewildered more than one historian. On some days Wilson appeared a southern Democrat, friend of laissez-faire, and unfeigned admirer of Edmund Burke; on others a crusader for a new world order, critic of laissez-faire, and ambitious tribune of the "new freedom." And while he played the conservative more noticeably before he went into politics, and championed "radical reconstruction" more forcefully after 1910, no period of his life was without some mixture of the "two Wilsons." Both his defenders and his critics, however, tend to overlook the persistent unity of his point of view. To his mind, Burkeanism and Progressivism seemed not in the least incompatible. "There is nothing so conservative of life as growth," he said at Princeton's sesquicentennial. "Progress is life, for the body politic as for the body natural. To stand still is to court death. Here then . . . you have the law of conservatism disclosed: it is a law of progress."[56]

Like Burke, though for different reasons, Wilson was suspicious of abstract principles—particularly natural rights—in politics, and was an opponent of the radicalism of the French Revolution. There is much in him that modern-day conservatives of a traditionalist stripe might like, though they'd regret it soon enough. He criticized Thomas Jefferson's "speculative philosophy," for instance, in terms that Russell Kirk could have approved, as "exotic" and "very aerated." Jefferson's theory "was un-American in being abstract, sentimental, rationalistic, rather than practical . . . ," Wilson charged. He remained a Progressive of a decidedly historicist and idealist type, not at all interested in understanding the past as it understood itself, much less recommending its revival. His appeal to Burke was rendered more plausible by the prevailing fashion of interpretation, which in the course of the nineteenth century had moved in utilitarian and evolutionist directions, well represented by Walter Bagehot and the so-called English Historical School. Again, one should pay careful attention to the thoroughly political context of Wilson's rhetorical appeals, and to his usual masterly control over language. He could sound Jeffersonian or libertarian, too, when he wanted. Much ink was spilled then and since over his statement to the New York Press Club in 1912 that "the history of liberty is the history of the limitation of governmental power." TR lambasted him for it. But Wilson never said that the *future* of liberty consisted in the limitation of governmental power. The future, he said repeatedly, would demand greatly expanded State powers, though not equally enlarged on every front.[57]

Power limited by all the familiar devices of American constitutionalism, from separation of powers to federalism, was a prescription for disaster, he maintained. Government "is not a machine but a living thing. . . . It is accountable to Darwin, not to Newton. It is modified by its environment, necessitated by its tasks, shaped to its functions by the sheer pressure of life." The competition of life demanded cooperation within each organism, so that it might constantly adapt to new external challenges. "No living thing can have its organs offset against each other as checks, and live," he concluded, very much in line with Hegel. Checks and balances make for a "fatal" warfare among the branches, rather than their "inti-mate, almost instinctive cooperation," which is essential to health. "*Living political constitutions*," therefore, "*must be Darwinian in structure and in practice.*" This "is not theory, but fact," he em-phasized, "and displays its force as fact, whatever theories may be thrown across its track."[58]

Though he may have borrowed the phrase from Bagehot, Wilson for all intents and purposes is the inventor of the theory of the "living constitution." It marks one of his two decisive contributions to the shaping of modern liberalism. Since the 1980s, the term has come up mostly in the discussion of judicial rulings and confirma-tion fights for liberal nominees to the federal courts. But Wilson ap-plied it not merely to the judiciary but to the whole political system, and to the executive more than the other branches. He was the first president to criticize the Constitution, to call it time-bound and incapable of meeting modern problems due to the very spirit and institutions of government that had once been its proudest boast. A living constitution had to be able to evolve, to change quickly its structure and function, to expand its powers to confront society's novel problems. Our eighteenth-century government had been de-signed, so Wilson charged, with the opposite end in mind. It sought to keep the Constitution as changeless as possible, and the times in tune with it; it managed the first, but not the second. As a result our evolving or, indeed, galloping social and economic problems had far outpaced any political solutions. What we needed, instead, was to keep the Constitution in tune with the times, or slightly ahead of them if possible, and for that either the old forms of gov-ernment had to be drastically amended, or the forms retained but an entirely new spirit breathed into them. As a young man, he pro-posed the former course, going so far as to suggest constitutional amendments to scrap the Constitution's separation of powers in favor of a British-style parliamentary system. That proved imprac-ticable, and he soon adopted the second strategy of new wine in old bottles. Plan B assigned to the president the crucial role of infusing the spirit of the living constitution into the old body of our laws and institutions. New kinds of presidential rhetoric and legislative agendas, new forms of expertise, regulation, and bureaucracy, and new Supreme Court justices and jurisprudence would be called for, all coordinated by the executive.[59]

Broadly speaking, this was the agenda of American liberalism over the next century, right up to the present day. Important new items would be added in the second and third waves of the liberal advance, but the fundamental alterations in the American *political* system were rooted in the turn from a "limited Constitution" (Hamilton's phrase) to a living one. Although Wilson presented it as an implication or application of Darwinism, substantively it owed even more to the theory of the State developed in nineteenth-century Germany. Hegel was the fountainhead of this State theory, but it had undergone several mutations by the time American universities began to flog students into studying it, and it would undergo more.

Before new wine could be put into the old bottles, the latter had to be emptied out. The living constitution implied a sweeping rejection of traditional American political principles, a point Wilson rarely could afford to make unambiguously, as we've seen. But if we briefly survey his own account in *The State*—his political science primer published in 1889, when he was a junior faculty member at Wesleyan—the departures from American orthodoxy will become clear. *The State* was part of the pioneering wave of American social science textbooks, designed to assist students who lacked German and Germanic attention spans.

First of all, the modern reader will be shocked to encounter the initial question raised in the book. What races shall we study? This question arose because Wilson relegated forms of government, once the heart of politics, to second or third place. The old political science had understood forms of government to embody partisan views of justice—who should rule, and for what ends. The choice among competing regimes had been the essence of politics, and of political science, an analysis still very much alive in the Declaration of Independence. But Wilson's assumption, following Hegel, was that there is never really much choice among regime types. Forms of government vary with the culture or folk mind, which varies with history, and thus political science is the story of the evolution from oriental despotism, to aristocracy, to democracy (Wilson thought that history had already left Hegel's constitutional monarchy behind). Rather than ever-present alternatives—even if some were less likely than others, depending on political conditions—the regimes were way stations in a one-way journey of political change. They "do not affect the essence of government" but "exhibit the stages of political development," to quote Wilson. As Fukuyama puts it today, liberal democracy is "the final form of human government," and except in very backward places in very marginal parts of the world there is no question of choosing any other form, certainly not in the long run. If you wish to encounter future princesses and lords, like Princess Leia and Lord Vader, you have to turn to science fiction.

But why focus on the races? Every world-historical culture was the possession of a people, a nation with its own peculiar religion, language, and ethnic or racial composition. History didn't choose the Greeks, the Romans, or the Germanic peoples randomly, after all, nor ignore the Africans capriciously. Each *Volk* was uniquely fitted by temperament, breeding, and even geography for the role it had to play. Here Social Darwinism entered into and modified the picture, by accentuating the racial or biological element that Hegel, though not all those inspired by him, had played down. (Not that he'd ignored it completely: African Negroes, he wrote, exhibit a lack of self-control and weak or nonexistent moral sentiments, while Germans display a becoming inwardness and "heart.") As a Progressive concept, *race* invited the promiscuous mixture of biological and cultural explanations precisely because it was tied to the idea of national evolution. The resulting notion of the State was "organic," all right, but in a new and illiberal way: the State was "the creation and expression of the racial community of the nation."[60]

Each "race" got the form of government it deserved, therefore, when it deserved it. And here is another unpleasant surprise for the modern reader. Wilson admitted that to be thorough, "every primitive tribe, whether Hottentot or Iroquois, Finn or Turk," ought to be studied. Practically, though, it was enough to know "the political history of the Greeks, the Latins, the Teutons, and the Celts principally, if not only, and the original political habits and ideas of the Aryan and Semitic races alone." A paragraph later he admitted that the Aryan race, by itself, would be his focus. "The existing governments of Europe and America furnish the dominating types of today," he explained. "To know other systems that are defeated or dead would aid only indirectly towards an understanding of those which are alive and triumphant, as the survived fittest." Political progress and the survival of the fittest precisely reinforced and confirmed each other. Did the "fittest" refer then to prowess in war or refinement in ethics, to *Macht* or *Recht*? The answer was that the two went together: Wilson presumed, without benefit of argument, that the societies that won their wars and survived their disasters, that were "alive and triumphant," were morally better than those "defeated or dead." But if the Nazis, ardent devotees of the Aryan if not the Semitic races, had triumphed in their war, or at least evaded total defeat, would that have proved them morally right? Of course not, and Wilson would have been among the first to agree that the fight was not over, or that this was no ethical advance because the "essentials" of the past had not been preserved. But these kinds of objections would suggest merely that his character was better than his principles. They would not extricate him from the problem that the Nazi constitution had passed the fundamental Darwinian test of survival, which he upheld as the *historical*, the "factual" standard of fitness. The bankruptcy of Progressive morality was never more than one lost war away.

Wilson had to face a preliminary version of the question after the outbreak of World War I. He had always tempered his Teutonic intellectual enthusiasms with a passionate love of British culture and parliamentary democracy, and had no trouble condemning, privately at first, the Kaiser's policy and form of government. TR was more open with his condemnation, despite the fact that he, and Wilson to a lesser extent, had spent decades extolling the achievements of the Teutonic, Aryan, and Anglo-Saxon races. After the bloody family feud broke out between the Teutons and the Anglo-Saxons, however, the Progressives, with a few exceptions, spoke far less amicably of their German ancestors. Besides, who exactly belonged to these races was always a question, especially in America with its increasing ethnic and racial heterogeneity. Hegel had written carefully of the Germanic, not merely the German, nations. But even at their most latitudinarian, the ranks of the Aryans, Teutons, and Anglo-Saxons in America did not include blacks, Irish, Chinese, and the "rabble," as John Burgess called them, of the Slavs, Czechs, Hungarians, and southern Italians. Invested heavily in the racial basis of politics, some Progressives undoubtedly sympathized with H. B. Adams's friend, the English historian Edward Augustus Freeman, who speculated that America's social problems might be solved if only " 'every Irishman [were to] kill a Negro and be hanged for it.' "[61] But the nature of American democracy did not permit Teutonic fantasies to keep politicians away from millions of immigrant voters for very long in any case, and the rabble who were initially fodder for the urban machines soon became fodder for Progressive uplift. This had stern and less stern modes, the former well represented by Roosevelt (no hyphenated Americans!) and the latter by, say, the social critic Horace Kallen ("cultural pluralism"). In many cases, however, the Progressives complicated the problem by imagining they were asking immigrants to subordinate their race to, or blend it with, the superior Anglo-Saxon one, rather than to exchange their previous political beliefs and habits for republican ones—as Jefferson and Adams, despite their disagreements, would have formulated the problem. The Founders viewed immigration as a question of citizenship, the Progressives as mainly a question of racial consciousness.[62]

For Wilson, the world dominance of the Aryan nations traced back over millennia to their ancient patriarchal family structure. Such families, "in which descent is traced to a common male ancestor, through a direct male line, and in which the authority of rule vests in the eldest living male," were either native to the Aryans or at least adopted early in the race's development. Non-Aryan families, by contrast, had a messy history of polygamy, polyandry, and promiscuity. Such disorder produced nothing but savage and stationary societies. Patriarchalism lifted the Aryans toward legitimate kingship as an early organizing principle, and gave them a leg up in the struggle for existence. Wilson's strange defense of patriarchalism as a political principle, albeit a primitive one, recalled its even more robust defense by Sir Robert Filmer, the seventeenth-century advocate of absolute monarchy best known as John Locke's whipping boy in the *Two Treatises of Government*. As a clue to just how hostile Wilson's political understanding was to the natural rights principles of the Constitution, this issue is extremely revealing. Filmer based the case for absolute monarchy on God's original grant of political power to Adam, which allegedly had descended ever since to the eldest males of the main extended families of mankind, conveniently organized into nations that ought to be headed by divine-right monarchs, who were conveniently in every case the heads of the existing royal families. It was one of the most absurd arguments in the history of political philosophy. Locke spent hundreds of pages refuting it, drawing a bright-line distinction between what he called "paternal power" and "political power." The one, the power of fathers over their families, had nothing to do with the other, the power that equal human beings entrusted to political officeholders to protect life, liberty, and property. The second kind of power was qualified, limited, voluntary, and revocable. The first kind applied within families—private societies held together by love rather than formal laws, and ruled by the presumed wisdom of the parents rather than by consent. The *parents*, note, not the father alone. Locke took great delight in reminding his readers that God had commanded "honor thy father and thy mother" equally, not honor thy father as the family godhead.

Family and political authority were thus two different things, just as civil society and civil government were distinct (though over-lapping) realms. Wilson affirmed the opposite, not only as a matter of historical record but as a matter of lingering justice. All government originated in kinship, he emphasized. "The original State was a Family. Historically the State of today may be regarded as in an important sense only an enlarged Family: 'State' is 'Family' writ large."[63] His purpose was not to bring back absolute monarchy, of course, but to clear the way for the comprehensive and compassionate authority of the modern State. Liberals ever since have cavalierly asserted that America is one big family, and that we are all our brothers' and sisters' keepers. President Obama is very fond of the brother's keeper appeal, using it countless times in his speeches.[64] We need not fear government's increasing power to be our keeper, according to liberals, because it operates as merely the most efficient instrument of our brotherly and sisterly duty to care for one another. Obama doesn't notice that God himself was pressing a narrow inquiry—in a missing persons turned murder case—when Cain replied with that famous rhetorical question, Am I my brother's keeper? God did not expect Cain to regulate Abel's shepherding or improve his diet, but to respect his brother's life and liberty. To that extent, at least, God was on the side of limited government. In a modern society numbering millions of inhabitants, the unwritten law of love will provide a watery bond, indeed, and government's efforts to enforce it will have to be out of all proportion to the intensity of "family" feeling. Hence the oleaginous compassion that has to be invoked and insisted on whenever government and family are confused. And why do liberals who regard the State as one big family never wonder how a family of all brothers and sisters is possible? Where are the parents in this metaphor? Who wields the adult discretion guaranteeing that the fearful authority of a national family—wielded by our big brothers and sisters who work for the State—will be used for the children's benefit? Like Obama, Wilson had no difficulty in seeing *himself* as the responsible grown-up in the room, and government as the wise father or at least the ever-present nanny, despite his insistence that "free men need no guardians."

Nor is the family relationship consensual. Children do not get to elect their parents, or to throw them out of office. Wilson's larger point was that government, based in kinship, was also originally an involuntary association. The Declaration had argued that all men are created equal, that as adults they are naturally free of each other's authority, and that they therefore had the right to choose their society and government with a view to their own safety and happiness. The Declaration's authors were far from believing that all or even many governments had ever begun under such happy auspices, of course. They were well aware that most regimes spring from "accident and force," in the words of *The Federalist*, not from "reflection and choice." But then they didn't think that history provided the standards for morality. The Progressive assumption, we have seen, is quite otherwise. "The facts of government mirror the principles of government in operation," as *The State* expressed it. "What government does must arise from what government is; and what government is must determine what government ought to do." Wilson had little patience, accordingly, for "the defects of the social compact theory" and set out to replace it almost entirely.

That theory was simply unhistorical. In primitive society, "the individual counted for nothing; society—the family, the tribe—counted for everything. Government came, so to say, before the individual." Any rights that individuals possessed came from their membership in the group, in society. There were no "individuals" by nature, hence no individual or natural rights, and hence neither social compact nor inherent limits on government. You could no more choose your government than you could your family, for the two were the same. To support his case against the free society Wilson turned to Aristotle, citing his dictum that "man is by nature a political animal," or rather revising it to say man is "a social animal." For though Aristotle's *Politics* had no precise equivalent of the "state of nature" in which unaffiliated human beings encounter each other as free and equal, he had sharply distinguished between the family and the city or political association, precisely on the ground (among others) of the freedom—supported by nature not society, and guided in the best case by reflection and choice—that men have to found new cities and reorder or revolutionize old ones. Wilson had to reinterpret Aristotle and the ancients to bring them in line with his historicist account of the growth of the State.

The Progressives' purported return to Aristotle was therefore vitiated from the start. It was impossible to return to Aristotle in the spirit of Darwin or via the premises of Hegel and the historical school. The attempt resulted only in the radicalization of the definition of "progress" as it moved farther and farther away from "the laws of nature and nature's God" as standards of political right. Man could not be *by nature* a political animal if his nature was itself a reflection of his society and its stage of historical development. That's why Richard Ely's translation or gloss of Aristotle was so telling: man is by nature not a political or even a social being, exactly, but a "State being." Wilson well understood that the nature of primitive government, the individual's complete submergence in society, was not the final word. Still, it was preserved as a "moment" in the fully developed democratic State of modern times. Through a historical sequence already explicated by Hegel and his disciples, the State had gradually come to cultivate, and at the same time to incorporate, the individual as such. "Men were no longer State fractions," Wilson wrote of the epic turn; "they had become State integers." What had begun as an unconscious project finally became a conscious one. The State, which had casually given birth to the individual, now dedicated itself to him. Compared to the ancient State, the modern one is "largely de-socialized," meaning it no longer absorbs the individual but "only serves him" by furthering his growth and self-realization. The State now existed for the individual, not the individual for the State. Yet these "State integers" still had no existence, and no proper rights or liberties, outside of the fostering State.[65]

What Wilson taught us to call a living constitution depended on this living State. The term, the *State*, was confusing then and remains so today. Its English language meaning, especially in the American context, suggested a part of some larger whole—a political subdivision of the Union, for example; or as a synonym for national government, the state as distinguished from society. Its German sense, though, was something else again. *Der Staat*, as Hegel had taught generations to conceive it, was the whole not the part, a synonym for the country or for the whole people politically formed. The State was the self-expression of the nation or *Volk*, "the organic body of society," as Wilson put it, without which the nation itself "would be hardly more than a mere abstraction." Not the result of a contract or compact, the State grew out of shared religious, cultural, ethnic, and linguistic identity, and necessarily expressed that identity. State and society were united, then, and the idea of the State as a necessary evil or as a potential threat to society was, from this point of view, absurd. They were the same organism, the same living people. It was only in and through the State that individual liberty and the moral life were possible. That, and the rational and ethical peak attained finally in the modern State, were what Hegel tried to convey when he pronounced the State "the image of God on earth." The German and English senses of the term were thus quite opposed to each other, and limited government made as little sense to the first as it made great good sense to the second. The confusion came from trying to make German political science (revealingly called *Staatswissenschaft*, the "science of the State") speak American English—or English speakers handle German political ideas. In any case, Wilson gave his heart to Hegel but his tongue to the American voter, whom he reassured that the modern State did not rule but "only serves."[66]

Wilson emphasized that with the historical emergence of the individual, the ancient idea of statecraft as soulcraft had been discarded. It's not the task of the democratic State to "prescribe the whole life of its citizens," to educate them to virtue. It must respect their freedom, their dignity as free and rational beings. But that implied not merely protecting their negative liberty to be secure in life, liberty, and property, but also their positive liberty. The latter involved not merely freedom *from* evils but freedom *for* the realization of all the goods, moral and intellectual, that were entailed by the State's promise of complete human development at the end of history. (If there were serious moral or intellectual desires left unsatisfied, history could not have ended yet.) In Wilson's words, "the individual must be assured the best means, the best and fullest opportunities, for complete self-development." Government's new purpose was partly to provide the conditions for that self-development, and partly to *equalize* the existing, unequal conditions that the capitalist economy provided to individuals and families. In other words, after sketching the huge distinction in principle that individual freedom made to the modern State, Wilson walked the argument back, stressing that it made only a very small difference in practice. For to fulfill its new mandate to provide or equalize the conditions for "complete self-development," in the name of freedom, not virtue, the State had assumed a vast portfolio of activities, nearly as vast as it had ever been but now harnessed to an unprecedentedly ambitious, and amorphous, goal. Wilson's State made Plato's *Republic* look modest. Plato's city had simply to educate every citizen's soul to its natural perfection as a member of the farmer, artisan, or guardian class. The means to that end were successively more fantastic, involving the noble lie, communism of property, women, and children, and finally the lucky emergence of philosopher-kings. Whereas Plato's scheme depended decisively on chance, Wilson's depended on the inevitability of primitive forms of life being superseded by advanced ones, or freedom realizing itself in the modern State. The overwhelming improbability and impracticability of the *Republic* was calculated to teach a profound moderation concerning a serious person's expectations of politics. Wilson's expectations of politics could hardly have been more extravagant. And promising or guaranteeing "complete self-development" as the common, harmonious goal of the individual, society, and the State—as the goal of the living constitution—was extravagance squared.[67]

Wilson tried his best to deny or at least postpone the utopianism implicit in this proposal. He sketched the paradoxical position that liberalism would cling to for the rest of the century: a seemingly libertarian indifference to the ends or purposes of human development, combined with a firm, moralistic insistence on government intervention in the economy. He couldn't specify what "complete self-development" would mean for every citizen or any citizen, or what the duty to protect and sponsor it would demand from government, except that no government regulation of "thought" or "conscience" or "private morals" would be permitted. Such interventions would violate freedom, and the whole point was to foster "the freest possible play of individual forces," an "infinite individual variety" of personal ends, to "aid the individual to the fullest and best possible realization of his individuality." He insisted, though, that "there must be constant adjustments of governmental assistance to the needs of a changing social and industrial organization." Though it is no longer master of the individual, government must reduce "the antagonism between self-development and social development to a minimum" and guard freedom "against the competition that kills," the unfair competition that creates "monopoly." He used that term loosely to mean a concentration of wealth and power in the hands of a few "selected by arbitrary fortune" rather than "by society itself." To those ends, he wrote in his own italics, ministrant government *does now whatever experience permits or the times demand*." One could without unfairness to him call this position individualism regarding the ends and socialism regarding the means, because he raised the issue of socialism himself.

> If the name had not been restricted to a single, narrow, extreme, and radically mistaken class of thinkers, we ought all to regard ourselves and to act as *socialists*, believers in the wholesomeness and beneficence of the body politic. . . . Every means, therefore, by which society may be perfected through the instrumentality of government, every means by which individual rights can be fitly adjusted and harmonized with public duties, by which individual self-development may be made at once to serve and to supplement social development, ought certainly to be diligently sought, and, when found, sedulously fostered by every friend of society. Such is the socialism to which every true lover of his kind ought to adhere with the full grip of every noble affection that is in him.

He accompanied this confession with his usual disclaimer, what he called "the rule of historical continuity." "In politics nothing radically novel may safely be attempted. No result of value can ever be reached in politics except through slow and gradual development,

the careful adaptations and nice modifications of growth." He believed that; but he also believed that something might appear to
almost everyone as "radically novel" until suddenly it didn't, that
is, until its success made clear it had been in preparation all along.
So it was with socialism, in the sense in which he used the term.
Socialism in the means gradually effects a socialism in the ends.
Self-development must in the end "serve" and "supplement" social
development.[68]

In *The New Freedom* he pictured the systematic renovation of
an old house, the republic under the Constitution, in which for a
generation or two we must continue to live until the day when "the
scaffolding will be taken away, and there will be the family in a great
building whose noble architecture will at last be disclosed, where
men can live as a single community, cooperative as in a perfected,
coordinated beehive. . . ." The hive will not be new; in fact by then
it will be exceedingly familiar, for we will have built it ourselves
without ever quite realizing it: the living Constitution is the hive
a-building. Only the kind of self-development that eventuates in a
political community "cooperative as in a perfected, coordinated beehive" will be sustained in the end, in other words. It may be that this
end will never quite be attained, if progress toward it remains asymptotic; but the beehive is the goal. The perfection of individuality must
go hand-in-hand with the perfection of sociality, because by "complete" self-development Wilson means not so much going as far as
an individual's talents or desires will take him, but the sort of growth
that rounds out all sides of the individual until his "public duties"
and "individual rights" become one. This is the "new freedom," the
right of perfect adjustment to the completely fulfilling State.

## Follow the Leader

But who gets to be queen bee? It's easy to forget that as a practical
proposal the living constitution does *not* imply allowing American government to adjust automatically to new social conditions or

problems. That kind of unplanned, natural response to new challenges would suggest something like natural selection, and would recall something like the virtues of the marketplace understood as an evolutionary testing ground. The Darwinism latent in the idea of the living constitution, however, is quite different. It's the Social Darwinism of artificial selection, of expert planning and experimentation. Someone must continually make decisions—launching studies, setting up new programs, expanding the Commerce Clause—to keep the living constitution alive, to adjust it to new circumstances. The term itself sounds almost green, like an environmental program the Republicans are always trying to kill. In reality, the living constitution is the liberals' constitution, even though they rule it not in their own name but in the name of their favorite experts, whose supposed expertise justifies their supposed authority to prescribe this and not that cure for our problems. Lester Ward, with his penchant for neologism, called the new form of government that would result from society consciously controlling its own evolution "sociocracy," the rule of society by society, but as usual he ignored the obvious point, which is that the effectual truth of the new regime would be the rule of society by sociologists.

Sociology bid fair to be the queen of the new social sciences. Influenced by the materialism and determinism of Social Darwinism, but also by sociology's own imperial claim, descending from its founder Auguste Comte, to have replaced philosophy as the highest and final form of human wisdom, the Progressives thought socioeconomic evolution to be the pacesetter of modern problems. Without supervision, society and the economy had developed rapidly but unhealthily in the nineteenth century. By putting aside laissez-faire and inaugurating rule by sociologists, the problems of modern life could be tamed, which meant forever managed by science. The problem of the age, after all, was that politics had systematically lagged behind economic and social evolution. So shouldn't the solution be to put the science of social evolution in charge?

Wilson went along with the diagnosis but not the prescription. He agreed that politics was part of the social organism, and that society had its own law of evolution. At times he sounded as though the only function left for politicians and political scientists was to play catch-up, to harmonize the political world with the social. He agreed that politics could not be (and never really had been) "architectonic" in Aristotle's sense, free to shape cities and souls according to the Legislator's or founder's view of justice. Politics had always been harnessed to the Spirit of the Age, though it had discovered it only recently. Although he conceded that politics could not rule society, he resisted the conclusion that it should merely reflect it. He rebelled against the notion that a cool process of scientific management would handle all social problems, that there was nothing great left to be done by citizens and statesmen. Human freedom and idealism must still count and count greatly, he insisted. And so in opposition to the spirit of scientism and to the exhaustion of noble striving implicit, as we have seen, in the end of history, he struck out in another direction. Politics could not rule society, but neither was it condemned meekly to follow it. Politics could lead society into the future.

Wilson was the first to celebrate "leadership" as an essential part of American democracy. Alongside the living constitution, leadership was the second element he contributed to the political definition of America's new liberalism. It's had a booming career. No one can run for president now, or even city council, without boasting of his or her leadership credentials. Nor has the idea's popularity been confined to politics. Modern business management is virtually unimaginable without every year's bestsellers on how to be an effective leader, how to inspire employees to follow the company's vision, and so forth. Nonetheless, some of those big-selling handbooks' titles—invoking everyone from Genghis Khan to Winston Churchill—suggest how much the concept owes to politics.

Leadership as such is neither a new term nor a new idea; it's been a highly visible part of political life. The ancient Greeks and Romans discussed the leader (*hegemon* and *dux*, respectively), and America's Founders did, too. Samuel Johnson's dictionary, the best one available to them, caught a certain contemporaneous ambivalence in the word: its first three meanings were straightforward (one who leads, or conducts; captain, commander; one who goes first) but the fourth ("one at the head of any party or faction") introduced a slight note of disapproval. That note grew louder in Johnson's addition to the definition, "as the detestable Wharton was the *leader* of the Whigs," and in his usage example, from Jonathan Swift: "The understandings of a senate are enslaved by three or four *leaders*, set to get or to keep employments." This distrust of factional or party leaders must have struck an American nerve. Unlike Johnson, the Americans were republicans with their own partisan prejudices against monarchical and aristocratic factions, but as far-seeing republicans they distrusted majority factions and popular leaders, too. Of the fourteen mentions of "leader" and "leaders" in *The Federalist*, for example, Madison and Hamilton use the terms pejoratively a dozen times, associating leaders with factions, vice, assaults on the law, and demagoguery in general. Jefferson and Adams shared this hatred of demagogues, and thus the suspicion of leaders. Stephen Douglas and Abraham Lincoln in their famous debates managed to get through twenty-one hours of oratory without a single discussion of their leadership abilities. The republican prejudice against leaders ran deep in the American political tradition.[69]

*The Federalist* did speak favorably of leadership twice, hailing "the leaders of the Revolution" and the people's "patriotic leaders" during the war. Those exceptions rather proved the informal rule. The revolution was a long war against the British Empire, and in war captaincy or the leadership of armies is necessary and proper, but as the examples of Caesar, Cromwell, and (soon) Napoleon showed, hardly without risk to republican freedom. Leadership as a phenomenon had a military cast, implying the use of force not reason, the imperative of issuing and following orders, not indulging debate, and thus inequality rather than equality of right and authority. A little bit of it went a long way in a free polity. George Washington was chief among the leaders of the revolution (Jefferson twice referred to him, in private letters, as "our great leader"), but for the rest of his life, as retired general and then president, he did everything he could to avoid being, or even being seen as, a party or factional leader. However much leadership may have been necessary to win wars and defend the Constitution against its enemies, before Wilson the concept seemed at odds with the rule of law, the priority of civilian to military habits and authority, and the equality of self-governing citizens that Americans associated with republicanism and democracy.

Most of these negative connotations have been forgotten. Leadership now seems a virtually unqualified good. It may be used for better or worse causes, to be sure, but in itself it suffers no disrepute or suspicion, despite the fact that one cannot be a leader without wishing to gain, and keep, followers. The mystery is why would a republic of self-governing citizens want to have leaders, which perforce would make them followers? It's not beyond the human capacity for self-delusion to imagine that in lusting after a leader the people are participating in leadership, or even taking turns at it. But there is something peculiarly democratic about imagining that we could all simultaneously and equally be leaders. In any event, how did democracy come to understand itself as relying on a steady supply of such men?

Wilson avoided the traditional objections by redefining leadership's purpose as something peaceful, rational, and beneficent: leading the people into the future. From a certain point of view, nothing could be more anodyne or self-evident. It was impossible to lead them anywhere else. By "the future," however, Wilson meant something grand, imminent, and if not quite knowable then at least vaguely imaginable: a world much better than the present, even as the present was much better than the past. Democratic leadership of this sort depended, in turn, on the presumptive truth of the Progressive philosophy of history, together with the Progressive refusal to declare history over quite yet. Because of the requisite gap between the present and the end of history, progress was still possible. But for the same reason, absolute knowledge wasn't yet in hand, though everyone was assured it was at hand. Political leadership rushed into the gap to help move society closer to its finale, but leaders couldn't offer absolute or scientific certainty about history's happy ending. They had to appeal to their followers not so much with arguments as with extrapolations or, since extrapolations didn't always inspire enthusiasm, with imaginative appeals concerning future marvels. Poetry had to come to the rescue of historical science. A new kind of idealistic or progressive political rhetoric was needed.

Wilson supplied that rhetoric in his own political speeches, which furnished the template for those of liberalism's other great presidential leaders, FDR, LBJ, and Obama, not to mention many lesser imitators. Wilson explained the model rhetoric in only one place, his speech-turned-essay "Leaders of Men." There he recounted the two new concepts that, as part of "leadership," would become the warp and woof of liberal political appeals ever since: vision and compassion. (He preferred to call the latter "sympathy," but these are synonyms divided only by their language of origin, Latin and Greek respectively.) With these three words he altered the American political vocabulary in fundamental and, so far at least, lasting ways.

He began by distinguishing the leader from the "literary man," for each is skilled at a form of "interpretation." The latter is a "sensitive seer" whose imagination can bring to life a thousand believable characters with motives not his own; his "subtle power of sympathy" affords him a "Shakespearean insight" into the hearts of individual men and women. The leader of *men* prefers to think in the plural, actually, in the mass. His sympathy is with what "lies waiting to be stirred" in the minds of "groups and masses," with a view to commanding them. He doesn't look up but ahead; he's guided not by the stars, by fixed reference points or principles of nature, but by the windings of the channel, by the great stream of time that has a flow and a direction that must be accommodated. Most leaders appear most of the time as compromisers, therefore, adjusting their measures to the prevailing circumstances, even as biological "growth is a process of compromise" between the organism and the environment. The leader must accommodate himself to the people he would lead, must interpret "the common thought" in order to determine "very circumspectly the *preparation* of the nation for the next move in the progress of politics." Society is an organism, and cannot safely be rushed; "slow modification and nice all-around adjustment" is the usual law of progress. Particularly in a democratic age, this law implies "that no reform may succeed for which the major thought of the nation is not prepared. . . ." Although "the ear of the leader must ring with the voices of the people," he's not a mere echo chamber. He has room for "initiative" but not "novelty," for "interpretation" but not "origination," all for the sake of encouraging "the tendencies that make for development."[70]

Yet progress doesn't always wear "the harness of compromise." "Once and again," Wilson wrote, "one of those great influences which we call a *Cause* arises in the midst of a nation." At that moment, history demands a more radical hope and change. Unharnessed, the leader is moved to stand for a "political or moral principle" that seems untimely, unpopular, almost unknown. The rallying cry of reform must then be presented as the leader's vision of the future—a vision of the turn history is about to take, his prophecy of the country's secular transformation. In this moment of radical freedom, of seeming liberation from the past, Progressive idealism swells to its full moral dimensions. The leader seems poised to remake the people after his own vision of them. Persuasion then functions as a kind of force. "There are men to be moved: how shall he move them?" Wilson asked. "He supplies the power; others supply only the materials upon which that power operates. . . . It is the power which dictates, dominates; the materials yield. Men are as clay in the hands of the consummate leader."[71]

It cannot be said that Wilson was innocent of the more ominous dimensions of leadership. This statement could serve as an eerie premonition (he gave the lecture in 1889 and 1890) of the leadership doctrines that emerged in Europe in the 1910s and 1920s, the ruthless, tyrannical, even nihilistic creeds of the communists and fascists. The fascination with leaders was not confined to America. As universal suffrage became the norm in more and more Western countries and mass democracy created mass political parties that needed to be coordinated and led, and as the old elites increasingly disqualified themselves for the job, the cry for leaders went up. Lenin responded, elaborating in *What Is To Be Done?* and other writings the "vanguard theory" that became Marxist-Leninist gospel. The workers could not be trusted to make a workers' revolution, he argued, because they were trapped in "trade union consciousness"—interested in better wages, shorter hours, more benefits, in everything except revolution. So the intellectuals and party bosses had to lead the revolution for them, acting as the proletariat's vanguard. By definition, elections would be futile to legitimate this vanguard, because the proletariat lacked sufficient proletarian consciousness. But the party apparatus could concentrate and project that consciousness, and under the leadership of the Central Committee of the Communist Party could force the people to make a revolution in the name of their future selves. Mussolini studied Lenin and the Bolshevik example. He switched from international to national socialism, downplayed class in favor of the national party, and the Central Committee in favor of the singular leader (*il Duce*)—and Stalin learned from Mussolini's example. Hitler streamlined the theory further, elevating race to the central fixation, and proclaiming *ein Volk, ein Reich, ein Fueh-rer*. Through the Nazi party, the racial consciousness of the whole nation would be drawn upward and concentrated in the person of the supreme leader. The Twenties and Thirties were terrible decades in which the dictators learned how to outdo each other in inhuman crimes and brazen tyranny, all in the name of leadership.

Fortunately for America, Wilson was a genuine democrat who kept his leadership theory firmly grounded in Progressive democracy. Although "men are as clay in the hands of the consummate leader," he isn't much of a potter, it turns out; he isn't concerned to mold them into anything radically new. The leader of men wields a "power which dictates, dominates"; for him the question concerns "the application of force." What he seeks, at least from one point of view, is merely to get the country moving again, moving toward the future, moving in line with its own innate, but temporarily stalled, law of historical development. A progressive people is not mere "materials," but matter with a certain inherent law of evolution or motion, which leadership can tap into.[72]

Individuals, though, are conscious primarily of their own attempts to move forward, which are mainly, though not exclusively, self-interested. Since the "forces of public thought" tend to be blind to so much, it's up to the leader to give them *vision*, in two senses: he has to improve their ability to see, to notice their fellows or the "community" moving alongside them; and he has to offer them a new object to be seen, a vision of the better future that is closer than they realize. To keep this vision democratic, Wilson insisted it had to remain intimately connected to the people. Sympathy or compassion, the leader's second new virtue, fits that bill. Leadership involves "sympathetic insight," or "a sympathy which is insight," into the nation's heart. As a part of the *Volk*, the leader must move with the common impulse and feel the common feeling. He must have "his rootage deep in the experiences and the consciousness of the ordinary mass of his fellow-men." But "drawing his sap from such sources," he can rise "above the level of the rest of mankind" and get "an outlook over their heads, seeing horizons which they are too submerged to see. . . ." Bill Clinton's "I feel your pain" (first said, even more unctuously, by Jimmy Carter) was the perfect expression of fellow-feeling advertising itself as a token of leadership. Compassion of this sort helps to keep the Wilsonian leader in touch with the led, with the community whose evolution he wishes simply to lead, not to dictate or overrule; it excludes Leninism, therefore; and it ensures the timeliness of reforms by binding the leader to his generation and its felt necessities. "No man thinking thoughts born out of time can succeed in leading his generation," Wilson wrote. Democratic leaders are merely "the more sensitive organs of society," whose keen vision lets them see on the horizon "some principle of equity or morality already *accepted* well-nigh universally" but not yet visible to the people themselves. Here is a thematic statement, taken from a description of Lincoln:

> A great nation is not led by a man who simply repeats the talk of the street-corners or the opinions of the newspapers. A nation is led by a man who hears more than those things; or who, rather, hearing those things, understands them better, unites them, puts them into a common meaning; speaks, not the rumors of the street, but a new principle for a new age; a man in whose ears the voices of the nation do not sound like accidental and discordant notes that come from the voice of a mob, but concurrent and concordant like the united voices of a chorus, whose many meanings, spoken by melodious tongues, unite in his understanding in a single meaning and reveal to him a single vision, so that he can speak what no man else knows, the common meaning of the common voice. Such is the man who leads a great, free, democratic nation.[73]

Leaders of a Cause, therefore, only seem out of touch with their times. They struggle to formulate and make explicit what lies "inchoate and vague" in the general sense of the community. At first society may resent them, refusing to answer their "rude and sudden summons from sleep." But once awakened, society will rise to the occasion and meet the "necessities of conduct" revealed to it. Accordingly, "no cause is born out of its time," he declared, and every reform—from the Protestant Reformation to the abolition of slavery—is "the destruction of an anomaly, the wiping out of an anachronism." This wasn't the way things looked to Lincoln or Luther, of course. And Wilson admitted that the timeliness of reform and the inevitability of victory in each case were obvious to "the judgment of history" or later historians, but not to the contemporaries themselves.

> Great reformers do not, indeed, observe times and circumstances. Theirs is not a service of opportunity. They have no thought for occasion, no capacity for compromise. But they are nonetheless produced by occasions. They are early vehicles of the Spirit of the Age. They are born of the very times that oppose them; their success is the acknowledgment of their legitimacy. . . . Is it not the judgment of history that [these reforms] were the products of a period, that there was laid upon their originators, not the gift of creation, but in a superior degree the gift of insight, the spirit of their age? It was theirs to hear the inarticulate voices that stir in the night-watches, apprising the lonely sentinel of what the day will bring forth.[74]

All statesmanship is a product of its age, but previous statesmen didn't understand this. They couldn't see the timeliness of their own reforms, the call of historical necessity that they answered. But "great reformers" all had an unconscious insight into the spirit of their age, a sense of what lay waiting to be developed, and a dominating passion to achieve something great. This led them to do history's bidding, all the while thinking they were doing their own bidding. Lincoln didn't realize he was destined to win, for example, or for that matter that winning would prove his legitimacy. But Wilson did—in his own case as well as in Lincoln's.

Without saying so, Wilson based his description of great reformers on Hegel's famous account in the *Philosophy of History* of "world-historical individuals." These were the men of great ambition and passion who swung history's gates, bringing one era to an end and commencing the next. Hegel's favorite examples were Alexander the Great, Caesar, and Napoleon—great slaughterers of men and destroyers of republics all, but whose actions answered not to ordinary morality but to the higher morality of history. "Many a gentle flower," remarked Hegel, had to be crushed by these monumental men and their destinies. They too acted on powerful instincts and insights into the needs of the age, and their motive was personal glory, not progress. Wilson's versions were more decent, democratic, and even religious, but served the same purpose on a smaller stage. History used their prodigious passions and unconscious instincts to bring about a more rational and moral world. But now that "the cunning of reason," as Hegel called it, had been exposed—now that great reformers knew they were tools of history, early vehicles of the *Zeitgeist*—what would democratic leadership amount to? What role could progressive leaders consciously play henceforth in American liberalism?

The succeeding princes of liberalism would provide the answer, but Wilson's own example was important. In the first place, leadership began to redefine the standards of American statesmanship, recasting them in a much more democratic mold. *The Federalist*, George Washington, and Lincoln had emphasized that one of the most vital if occasional tasks of the president would be to defend the Constitution and sound laws and policies against the temporary delusions and passions of the people. The populist nature of Wilsonian leadership began almost immediately to marginalize that duty and indeed that whole conception of fidelity to constitutional forms and their foundation principles. Into their place stepped the leader's sympathy with the people, his reliance on them as his indispensable connection to the Spirit of the Age. "I have often thought," Wilson said while campaigning in 1912, "that the only strength of a public man consisted in the number of persons who agreed with him; and that the only strength that any man can boast of and be proud of is that great bodies of his fellow citizens trust him and are ready to follow him." If anything, the second clause is more cringe-worthy than the first, a complete abdication of all standards of internal probity and external principle in public life. "For the business of every leader of government," Wilson said elsewhere the same year, "is to hear what the nation is saying and to know what the nation is enduring. It is not his business to judge *for* the nation, but to judge *through* the nation as its spokesman and voice." The descent from magnanimity to pusillanimity as the standard of high public officials was predicted in that formula. No longer would a statesman conceive of himself as offering the public his best judgment of their interests, and his service to their opinions so long as it was honorable to do so (that had been Burke's standard). No longer, even, would leading the nation be thought to require from the leader a certain commanding or world-historical height, "a new place of outlook and of insight." In the modern State as it developed to maturity, it would be enough for the leader to judge "*through* the people" and to act as their "spokesman and voice." The statesman as ventriloquist—that was where the doctrine of leadership was headed. To be a leader, it eventually would suffice to be merely in the lead, a little out in front of the people, an early-adopter, the kind of visionary for whom it is vital to stay a few steps ahead of the people, the kind of visionary who has a very good pollster. The secret of leadership would be revealed to be good followership.[75]

Or perhaps these worries would prove superficial, because the leader would mollify the people with pleasant words but reserve a private sanctum for the vision that speaks to him personally. That could produce a danger of a more sinister, more acute kind: the leader with his own vision of what President Obama calls "absolute truth." *Vision* is a term, and a phenomenon, with a long religious history, and Wilson surely had its biblical usage in mind. As God sent a vision to his prophets, so history sends a vision to the leader of men. But Wilson tried to keep the people in the divine loop, and the leader closely bound to them. But what kind of a leader is always in leading-strings? Doesn't he ever want to stand on his own? Anticipating the problem, the authors of *The Federalist* condemned "visionary" politicians or schemes as disasters waiting to happen.[76] To his credit, Wilson recognized the theoretical danger, as it were; but he didn't fear it as a practical danger in an Aryan country with a centuries-old tradition of self-government. He assumed that both the people and their leaders had inherited the acquired characteristics of Anglo-Saxon self-rule. All of those vices of human nature that had doomed democratic governments and peoples in the past, and that the Framers had tried to guard against in the Constitution, had been overcome, Wilson concluded.

At the beginning of the great crisis of his time, Lincoln warned in 1854:

> Slavery is founded in the selfishness of man's nature—
> opposition to it in his love of justice. These principles are
> in *eternal* antagonism; and when brought into collision
> so fiercely as slavery extension brings them, shocks, and
> throes, and convulsions must ceaselessly follow. Repeal the
> Missouri Compromise—repeal all compromises—repeal
> the Declaration of Independence—repeal all past history,
> you still cannot repeal human nature. [Emphasis added.]

Lincoln thought, in the words of his great interpreter Harry V. Jaffa, "that the principles of political freedom were grounded in nature and eternity, not in history." Although Wilson lived to see the opening horrors of the twentieth century—the World War and the Bolshevik Revolution—he clung to his faith that the antagonisms in human nature were being overcome, that selfishness would eventually, inevitably yield to the love of justice. It was therefore safe to indulge a politics of vision, and alongside it a politics of the living constitution, that took their guidance from history, even though history was not over yet. Great reformers are "born of the very times that oppose them; their success is the acknowledgment of their legitimacy," Wilson wrote. If you win, he taught in effect, you must be right; or at best, if you are right, you must eventually win. In practice, the moral lesson was the same: justice equals success, the right of the stronger. He didn't shy away from that moral. "I would a great deal rather lose in a cause that I know someday will triumph," he told the voters in 1912, "than triumph in a cause that I know someday will lose."[77] It was not even necessary to add that the cause which "someday will triumph" would be a just one. Here was the perfect fusion of secularized Christianity and Social Darwinism; and the perfect confusion of the triumph of justice with the triumph of the will. Deeply wrought in Progressivism, this confusion of morality with history and this blindness to human nature and natural right would haunt the subsequent waves of American liberalism.

# 3

# Franklin D. Roosevelt and the Rise of Liberalism

In the euphoric weeks after the 2008 election, liberal commentators congratulated President-elect Obama on a victory so sweeping that it invited comparison to Franklin Roosevelt's in 1932. *Time* magazine's cover proclaimed "The New New Deal," though the headline seemed redundant against the cover art of FDR at the wheel of a 1930s roadster, with his cigarette holder clutched jauntily in his teeth—make that Obama's teeth, his face digitally switched, the only incongruity being FDR's too-small gray fedora resting on the incoming president's big, brainy head. The point was not simply that both were smokers. As Peter Beinart explained in the cover story, "The New Liberal Order," if Obama "can do what FDR did—make American capitalism stabler and less savage—he will establish a Democratic majority that dominates U.S. politics for a generation."[1]

The two achievements were connected, taming capitalism and building an enduring Democratic majority. No one understands this better than Obama himself. "The last time we faced an economic transformation as disruptive as the one we face today," he wrote in

*The Audacity of Hope*, "FDR led the nation to a new social compact—a bargain between government, business, and workers that resulted in widespread prosperity and economic security for more than
fifty years." At the National Press Club in 2005, Obama explained
that Roosevelt "understood that the freedom to pursue our own individual dreams is made possible by the promise that if fate causes us
to stumble or fall, our larger American family will be there to lift us
up. That if we're willing to share even a small amount of life's risks
and rewards with each other, then we'll all have the chance to make
the most of our God-given potential." Risks and rewards—the new
social contract was not merely social insurance designed to spread
the risk, but a form of income redistribution or wealth-sharing as
well. His remarks in 2005 were directed against President George
W. Bush's plans to reform Social Security by grandfathering in individual retirement accounts, by privatizing (even if only partially)
the program established by FDR in 1935. It might seem strange that
a self-described progressive, of all people, would bitterly resist the
idea that after some seventy years a government program might need
fundamental reform, or at least an adjustment or two to meet new
realities. He admitted that the Bush administration pointed to "how
much things are different and times have changed since Roosevelt's
day." But the only changes Obama could see were for the worse.
Although America isn't suffering "the absolute deprivation of the
Great Depression," our relative deprivation and insecurity are as bad
as, or worse than, they were then. Despite the enormous increase in
national wealth and the standard of living, we need Social Security
now more than ever, and a host of new social welfare programs,
too—federal support for child care, universal health insurance, and
more. The personal retirement accounts (combined with reduced but
still guaranteed Social Security benefits) offered by the Republicans
would amount to—Social Darwinism![2]

What Obama feared was that one act of privatization would
lead to another, until the new (now old) social compact envisioned

by "the genius of Roosevelt" would unravel. There would eventually be a time for "the spirit of pragmatism and innovation" to be brought to bear on the problems of Social Security, he reassured his audience, but now was not it; and anyway, the spirit of Bush's reform was "purely ideological," not pragmatic. How curious then that Obama's objection to it was not that the plan wouldn't work, the quintessential pragmatic objection, but that it would work all too well to accomplish its nefarious end. At the heart of his revulsion to Bush's proposal was the ideological or moral concern that it would strike at the "comfort and dignity" offered to Americans by the New Deal program and at the "basic standard of living" enshrined in the social compact underlying the modern welfare state.[3] Obama also had in mind, of course, an objection that neatly combined the pragmatic and the ideological, namely that any hope of establishing a new "liberal hegemony" that would "hold sway in Washington until Sasha and Malia have kids" (to borrow Beinart's language) would be dashed if FDR's social compact were allowed to be significantly revised or repudiated.

## From Progressivism to Liberalism

Franklin Roosevelt was the second great captain of liberalism, the first to fight under that banner and the first to win electoral victories so massive and legislative successes so lasting that liberal public policies and, even more important, the assumptions behind those policies, became ruling elements in our public life. By every standard but one, his political achievements were far greater than Wilson's. To Wilson belonged credit for discovering and exploring the new continent of American politics; FDR colonized and held it. As a young man, FDR had been an ardent Progressive. He married a niece of TR's, and served as assistant secretary of the navy in the Wilson administration. His first biographer, Ernest K. Lindley, who later wrote the occasional speech for FDR, considered him steeped in the ideas of the New Freedom, though "he distrusted all economic theories."[4] In his acceptance speech at the 1932 Democratic National Convention, FDR strikingly described his relation to his old boss, who had passed away eight years earlier. Calling for the resumption of the nation's "interrupted march along the path of real progress," Roosevelt declared that "our indomitable leader in that interrupted march is no longer with us, but there still survives today his spirit. . . . Let us feel that in everything we do there still lives with us . . . the great, indomitable, unquenchable, progressive soul of our Commander-in-chief, Woodrow Wilson." A few months later, he explained that "had there been no World War— had Mr. Wilson been able to devote eight years to domestic instead of to international affairs," then the problem of our "highly centralized economic system, the despot of the twentieth century," might have been solved or at least mitigated. By FDR's own argument, the New Freedom's incompleteness set the problem and pointed to the goal of the New Deal, which would enjoy, as it turned out, two full terms to devote to domestic reform before the gathering storm of world war would again force a shift to international affairs.[5]

To many historians and biographers, of course, Roosevelt was hardly worth taking seriously as a thinker. Many of his contemporaries agreed. Oliver Wendell Holmes Jr., the eternally serving Supreme Court justice appointed by Teddy Roosevelt, set the pattern with his famous remark that FDR had "a second-class intellect but a first-class temperament." Two of his closest advisors, Raymond Moley and Rexford Tugwell, spoke in retrospect of the "struggle for Roosevelt's mind." According to Richard Hofstadter FDR was an "opportunist," groping his way through eight years of trial-and-error reforms with no consistent purpose or intention at all. A later admirer, James MacGregor Burns, agreed that "Roosevelt was no theorist. It is doubtful that he chose this course as a result of a well-defined political philosophy. It simply emerged." Other historians damned by faint praise. Though "not an intellectual," Paul Conkin wrote, "Roosevelt was neither an anti-intellectual nor a pseudo-intellectual. He had no pretensions as a thinker."[6] But it's possible to be an unpretentious thinker, one who doesn't call attention to his thought because it's not in his or his thought's interest to do so. As Tugwell noted, FDR had studied Wilson's example and had learned

> that to furnish a blueprint for progressive friends is to furnish one also for reactionary enemies; and when it is known that most friends are demanding it and will denounce it bitterly in any case because its center of reference differs from their idea of what it should be, and when enemies control all approaches to the public mind and all the accumulated wealth of the country, a wise man will think twice before he "telegraphs his next punch," to say nothing of publicizing a whole plan of campaign.[7]

Perhaps the most ambitious thinker-doers wielding executive power would prefer to attribute their choices to necessities that leave them no choice.

At any rate, we know that nearly all FDR's speeches were drafted by advisors, but we know also that he demanded rewrites until he got what he wanted, often editing the drafts himself at the end. The political scientist Robert Eden has made a careful study of the drafting of one of Roosevelt's most significant addresses, to the Commonwealth Club in San Francisco in 1932, and found evidence of the then-candidate's pervasive influence over its theme and composition.[8] FDR was no Wilson in terms of his publications and intellectual record, but then he needn't be. The fundamental intellectual work in the broadest sense had been done, and the vaunted members of FDR's Brains Trust, for all their luster, were policy experts and good writers rather than theorists, though Adolf Berle and Tugwell came close. At any rate, Roosevelt himself was keenly aware of the inescapable educational dimension of democratic statesmanship. As he stated to the Commonwealth Club, "Government includes the art of formulating a policy, and using the political technique to attain so much of that policy as will receive general support; persuading, leading, sacrificing, teaching always, because the greatest duty of a statesman is to educate."

Roosevelt's own political genius may have owed more to insight than reflection, but it has been underrated regardless. One has to remember that by 1918 or 1920 at the latest, the air had gone out of the Progressive balloon. The Senate's refusal to ratify the Treaty of Versailles; America's self-exclusion from the League of Nations; Wilson's stroke and the crippled presidency that followed, presided over by his wife, Edith; recoil at the costs of "war socialism" to free speech and a free economy; and above all disillusion with the world war itself, its murderous death toll, the vindictive peace, and the idealism consistently invoked by all sides on its behalf—these did not bespeak progress, much less the inevitable progress in human power and morals promised by the movement bearing the name.[9] To add insult to injury, the 1920s, dominated by the Republican Party and the return to normalcy, proved staggeringly prosperous. For the Progressives, these were wilderness years, which did not look likely to end soon. John Dewey and others, despairing of the GOP's future and the Democratic Party's past and present, contemplated a third party for true believers who could resuscitate Progressivism and gradually undermine America's capitalist culture. In a 1926 editorial, William Allen White caught the mood of the times.

> The spirit of our democracy has turned away from the things of the spirit, got its share of the patrimony ruthlessly and gone out and lived riotously and ended by feeding among the swine. Perhaps if [TR] had lived, he would not have permitted public sentiment to sag as it has sagged. . . . The nation has not yet been shocked out of its materialism.

Of course, Coolidge is a tremendous shock absorber. His emotionless attitude is an anaesthetic to a possible national conviction of sin. . . . We have just to grind along and develop a leader and it is a long, slow task calling for all our patience. How long, Oh Lord, how long![10]

When the Depression hit, many of the Progressives assumed, probably as did most Americans, that like previous slumps it would be over in a few years, the boom would resume, and nothing major would have changed in our politics, except perhaps that many of the social reforms won in the preceding two decades would be swept away as unaffordable and counterproductive.

Franklin Roosevelt boldly seized the moment to redefine the Depression as, in Arthur M. Schlesinger Jr.'s phrase, the "crisis of the old order," and to prescribe as the cure a Democratic Party reborn as the party of modern *liberalism*. FDR never eschewed the label *progressive* (though usually small-*p* progressive), but over time he leaned more and more heavily on *liberal*. (There was also the awkward fact that Herbert Hoover was well-known as a Progressive himself, having voted for TR and served in the Wilson administration.) After saluting Woodrow Wilson's "progressive soul" in the 1932 acceptance speech, FDR stated some of the party's ideals, beginning with "the fact that Democratic party by tradition and by the continuing logic of history, past and present, is the bearer of liberalism. And of progress and at the same time of safety to our institutions." A few paragraphs later he proclaimed, "Ours must be a party of liberal thought, of planned action, of enlightened international outlook, and of the greatest good to the greatest number of our citizens." He titled the 1938 volume of *The Public Papers and Addresses of Franklin D. Roosevelt* (edited by his old friend Samuel I. Rosenman) "The Continuing Struggle for Liberalism." In that volume, which covered his so-called purge campaign against reactionary Democrats in the primaries, he devoted a large part of his introduction, in addition to six speeches reprinted in the collection itself, to the meaning of liberalism. He embraced the word with fervor. "I believe it to be my sworn duty, as President, to take all steps necessary to insure the continuance of liberalism in our government," he wrote. "I believe, at the same time, that it is my duty as the head of the Democratic party to see to it that my party remains the truly liberal party in the political life of America." It's revealing that he felt he had a *constitutional* duty to make and keep the government liberal. As he explained, liberalism and conservatism are usually the "two general schools of political belief" in a representative government. The liberal party—which took different names at different times in our history—stood for two things: "the wisdom and efficacy of the will of the great majority of the people," and the duty of government to "use all its power and resources to meet new social problems with new social controls—to insure to the average person the right to his own economic and political life, liberty, and the pursuit of happiness." Majoritarianism and the living constitution—that was his liberalism. His interventions in the 1938 party primaries stemmed from his conviction "that my own party can succeed at the polls only so long as it continues to be the party of militant liberalism." Putting the best face on Democratic losses (the first net Democratic losses, remarkably, since 1928—though that still left the Democrats with 69 senators and 262 representatives, versus 23 Republican senators and 169 Republican congressmen) in the elections that year, he concluded, "Liberalism in Government was still triumphant."[11]

The embrace of the new label was plain. He was aware, of course, that liberalism had an older meaning and tradition, what we would loosely call nowadays "classical liberalism," embracing individualism, free markets, and limited government in the various ways they had been championed since the seventeenth and eighteenth centuries. He had had occasion to be reminded of the older definitions by Herbert Hoover, who claimed to be the true liberal and had tried to vindicate his claim in his 1934 book, *The Challenge to Liberty*. "Liberalism holds that man is master of the state, not the servant; that the sole purpose of government is to nurture and assure these liberties," wrote Hoover. "All others insist that Liberty is not a God-given right; that the state is the master of the man. . . ." Two years later he told Republicans, "The party must become the true liberal party of America." The New Dealers represented a "false liberalism" that would "limit human freedom and stagnate the human soul." Their ideas, he said that same year, "are dipped from cauldrons of European fascism or socialism."[12]

Until this debate broke out in the 1930s, liberalism had not been an important term of distinction in American politics. A rump party calling itself the Liberal Republicans had contested the presidency, once, in 1872. A few groups opposing Prohibition had used the name. Within Progressive ranks, "liberalism" suggested laissez-faire, a bête noire to the reformers. Herbert Croly and his colleagues at the *New Republic* were perhaps the first to adopt the term as something positive, beginning around 1915 to identify themselves with "the cause of liberalism." Their candidate, TR, had been defeated in 1912, and though they endorsed Wilson in 1916 the old vocabulary of reform and Progressivism had perhaps lost its attraction. The war, too, may have moved them away from German concepts central to Progressivism, such as the State, and toward English or Anglo-American ones, such as what George Santayana, who published three articles in 1915 on the Anglo-German comparison, called "the liberty of liberalism." Croly and many others admired the Liberal Party of Great Britain, which had pioneered a new, socially conscious liberalism for decades. Certainly issues raised by the prosecution of the war—academic freedom, conscription, censorship—may have had an effect on the magazine's new appreciation for individual freedom. But the *New Republic*'s rehabilitation of the term did not catch on in politics at large. John Dewey drew attention to the concept in a series of articles and books discussing the meaning and history of liberalism, pitting its humanitarian or liberating ends against its now-obsolete means of limited government and capitalist economics—which means had to be rejected so that a revived liberalism could flourish.[13] When all is said and done, however, it was still FDR who firmly, enduringly applied the liberal label to his cause and to his party.

Why did he make the change? Certainly he wanted to put some daylight between himself and the "Progressivism" of a failed third party.[14] Plus he liked to discomfit his enemies. One advantage of the maneuver was to deprive his opponents of a home, rhetorically speaking. It left them protesting—see Hoover, above—that they'd been robbed. Having lost their good name, "the immediate jewel of their souls," what were they supposed to call themselves? FDR suggested, helpfully, that they ought to call themselves conservatives, a designation they were loath to accept because it sounded reactive, empty (conserve what, exactly?), and vaguely un-American, that is, Tory. It took a long time for conservatives to warm to the term; Robert Taft, "Mr. Conservative," was still insisting he was a liberal in 1946. Barry Goldwater, in 1963, told young Republicans to beware "of the phoniness that has been going on under the false guise of liberalism for the last thirty years."[15] Yet the greatest advantage FDR reaped from the new nomenclature was that it made it easier to connect to America's past and particularly to America's Founders. Wilson had been of two minds about the Founders. He was prepared to claim them as his ancestors, insofar as they had been "practical statesmen with an experienced eye for affairs and a quick practical sagacity"; but insofar as they were Newtonians and "Whig theorists," he had to condemn their erroneous and influential doctrines. Accordingly, the New Freedom was explicitly, in contradistinction to the old, a new doctrine for a new time; it was precisely not "a new birth of freedom" in Lincoln's sense, that is, a spiritual rebirth and deepening of the old.[16] FDR's liberalism was calculated to acknowledge and keep open a bridge leading back to the Founders and, incidentally, to Lincoln. Much more plausibly and much less ambivalently than Wilson, Roosevelt intended to claim for the Democrats and for his own administration the title deeds of the American Founding. At the same time, he intended to read the Republicans—or the Republican leaders, a distinction he insisted on—out of the American political tradition or at least out of its mainstream consensus, casting them into the outer, undemocratic darkness of fascism, communism, and Toryism.

He went at this larger purge with great zeal and effectiveness. In his acceptance speech in 1932 he fired the initial volley.

> There are two ways of viewing the government's duty in matters affecting economic and social life. The first sees to it that a favored few are helped and hopes that some of their prosperity will leak through, sift through, to labor, to the farmer, to the small-businessman. That theory belongs to the party of Toryism, and I had hoped that most of the Tories left this country in 1776. But it is not and never will be the theory of the Democratic party.

This early statement of trickle-down economics condemned it not merely as bad economics but as downright un-American. Four years later, FDR arranged that the Democratic Party platform be written as a loose imitation of the Declaration of Independence. The platform began with five paragraphs, and ended with another paragraph, boasting self-evident truths. Here is a sample of the beginning, along with the concluding passage.

> *We hold this truth to be self-evident*—that the test of a representative government is its ability to promote the safety and happiness of the people.

*We hold this truth to be self-evident*—that twelve years
of Republican leadership left our nation sorely stricken in
body, mind, and spirit; and that three years of Democratic
leadership have put it back on the road to restored health
and prosperity.

*We hold this truth to be self-evident*—that twelve years
of Republican surrender to the dictatorship of a privileged
few have been supplanted by a Democratic leadership
which has returned the people themselves to the place of
authority, and has revived in them new faith and restored
the hope which they had almost lost.

. . .

*We hold this final truth to be self-evident*—that the interests, the security and the happiness of the people of the
United States of America can be perpetuated only under
the democratic government as conceived by the founders of
our nation.

The last phrase was masterly: the democratic government (read:
Democratic government) as conceived (an echo of the Gettysburg
Address, but with conceived "in liberty" changed, in effect, to in
democracy) by the Founders, who, if present today, he implied,
would be subscribing to this new Declaration of Independence
from the Republican Party's plutocracy and dictatorship.[17]

Roosevelt followed up on the platform's language with his most
sweeping denunciation yet of the enemy within, in his acceptance
speech at the Democratic National Convention, which was meeting in Philadelphia, on June 27, 1936, only a week away from the
160th anniversary of the Declaration's signing in the same city.

Philadelphia is a good city in which to write American
history. This is fitting ground on which to reaffirm the faith
of our fathers; to pledge ourselves to restore to the people

a wider freedom; to give to 1936 as the founders gave to
1776—an American way of life.

His reaffirmation and restoration were also, therefore, part of a
larger practical effort to re-found ("as the founders gave to 1776")
the American way of life. He continued:

> In 1776 we sought freedom from the tyranny of a politi-
> cal autocracy—from the eighteenth century royalists who
> held special privileges from the crown. . . . Political tyranny
> was wiped out at Philadelphia on July 4, 1776. Since that
> struggle, however, man's inventive genius released new
> forces in our land which reordered the lives of our people.
> The age of machinery, of railroads; of steam and electric-
> ity; the telegraph and the radio; mass production, mass
> distribution—all of these combined to bring forward a
> new civilization and with it a new problem for those who
> sought to remain free.
>     For out of this modern civilization economic royal-
> ists carved new dynasties. New kingdoms were built upon
> concentration of control over material things. Through new
> uses of corporations, banks, and securities, new machinery
> of industry and agriculture, of labor and capital—all un-
> dreamed of by the fathers—the whole structure of modern
> life was impressed into this royal service.

These powerful lines reveal why reaffirmation and restoration
were not enough. The "whole structure of modern life" was "un-
dreamed of by the fathers," so we must emulate but not repeat
their creative statesmanship. They *solved* their age's problem—political tyranny, the denial of government by consent and basic civil
rights—which in America we no longer need fear. In modern times
a new disease, "economic tyranny," attacked our civil rights only

by means of a secondary infection; its real target was our newly emergent economic rights.

> It was natural and perhaps human that the privileged princes of these new economic dynasties, thirsting for power, reached out for control over government itself. They created a new despotism and wrapped it in the robes of legal sanction. In its service new mercenaries sought to regiment the people, their labor, and their property. And as a result the average man once more confronts the problem that faced the Minute Man.
>
> . . . For too many of us the political equality we once had won was meaningless in the face of economic inequality. . . . For too many of us life was no longer free; liberty no longer real; men could no longer follow the pursuit of happiness.

A peculiarity of the speech was that it never mentioned the Republican Party or any Republican leader by name, but this fact only confirmed the force of FDR's indictment. Everyone knew who the economic royalists were.[18]

The Democrats' political victories built on one another, increasing their majorities in Congress and their popular vote for president from 1932 until the wave broke in 1938. They ruled by such supermajorities that believers in the living constitution then and now came to regard the New Deal congresses as a sitting constitutional convention, altering and adjusting the Constitution at will, informally yet decisively.[19] Yet Roosevelt still feared a conservative counterrevolution, and warned against it. "Too many of those who prate about saving democracy are really only interested in saving things as they were," he said in 1938. "Democracy should concern itself also with things as they ought to be." Then he lowered the boom. "I reject the merely negative purposes proposed by old-line

Republicans and Communists alike—for they are people whose only purpose is to survive against any other Fascist threat than their own." The breathtaking parallelism of "old-line Republicans," of men like Coolidge and Hoover, with communists, and the even more reckless equation of Republicans with fascists, was not, alas, an aberration. It would become standard fare in FDR's ruthless effort to whip his opponents into line in support of, or at least in acquiescence to, New Deal policies and principles—to make the New Deal permanent, to raise it to constitutional status, effectively beyond the reach of future political majorities. He went on:

> As of today, Fascism and Communism—and old-line Tory Republicanism—are not threats to the continuation of our form of government. But I venture the challenging statement that if American democracy ceases to move forward as a living force, seeking night and day by peaceful means to better the lot of our citizens, then Fascism and Communism, aided, unconsciously perhaps, by old-line Tory Republicanism, will grow in strength in our land.

In this passage he was not speaking of GOP isolationism or views on foreign policy, but of its allegedly "Tory" resistance to the New Deal domestic agenda. He accused the Republicans of being devotees of dictatorship and minority rule, alongside the fascists and communists, whom the Republicans aided, "unconsciously perhaps."[20]

Roosevelt's insistence that the New Deal was the only viable middle way between right-wing and left-wing extremism, the only possible response to fascism and communism, was always an unlovely aspect of his politics. In 1932, at the bleak bottom of the Depression, the Socialist Party candidate for president got a whopping 2 percent of the popular vote, and the Communist Party candidate one-quarter of 1 percent. This was the imminent threat to the American way of life? But in later years, as his political strength waned, he grew even more insistent that the New Deal was the sine qua non of democracy and of morality. This is from his last annual message to Congress, in which he proclaimed (as we shall see) the Second Bill of Rights, in 1944:

> One of the great American industrialists of our day . . . recently emphasized the grave dangers of "rightist reaction" in this Nation. All clear-thinking business men share his concern. Indeed, if such reaction should develop—if history were to repeat itself and we were to return to the so-called "normalcy" of the 1920s—then it is certain that even though we shall have conquered our enemies on the battle-fields abroad, we shall have yielded to the spirit of fascism here at home.

Harding, Coolidge, Hoover, Andrew Mellon—liberals all, in the old sense of the term, but now it seemed no better than Hitler, Goebbels, Himmler, and Ribbentrop. Politicians often say hyperbolic things, of course, and FDR probably smarted from many of the GOP's full-throated denunciations of him. But coming from the commander in chief in wartime, in a State of the Union message, without restraint or equivocation, without any Jeffersonian or Lincolnian demurrals about every difference of opinion not being a difference of principle, or a common loyalty to the Constitution binding all parties despite their vigorous disagreements—and as part of a campaign to promote and legitimate a new kinder, gentler liberalism, well, it was an unforgettable example of the ruthlessness of the pure in heart.

Roosevelt's decision to fight under the banner of liberalism was crucial, then, to his ability to define others as illiberal, to reconstruct the American political tradition so that it culminated in the New Deal, and to execute an electoral realignment that would make the Democrats the majority party for several generations. For these plans to work, the people had to go along step by step, and they did. They choked on the Court-packing plan, and FDR worried for a moment that he would lose them, but they returned to him easily. In part that was because he had persuaded them that faith in his leadership meant keeping faith with the Declaration of Independence and the Constitution, for he had succeeded in winning the power, and thus the right, to interpret them for his time. His conception of himself as a leader, with a vision of the future drawn from the people themselves, was straight out of Wilson, as was the purported need to reinterpret those American scriptures to keep them alive and timely.

But liberalism had another side for Roosevelt, one that expanded on Wilson's exaltation of "sympathy." Hoover and FDR would both have embraced the original sense of liberalism as pertaining to *homo liber*, the free man who is not a slave. They disagreed, among other things, on whether man finds his security in freedom, or man finds his freedom in security. They would have acknowledged also the connection between liberalism and *liberalitas*, the virtue of generosity, liberality, kindness. Hoover regarded this as primarily if not exclusively a private virtue, at home in civil society. Roosevelt insisted it was a public and governmental virtue, without which government couldn't be, as it should be, the self-expression of a compassionate people. "We seek not merely to make government a mechanical implement," he said at the 1936 party convention, "but to give it the vibrant personal character that is the very embodiment of human charity. . . . Governments can err, Presidents do make mistakes, but the immortal Dante tells us that divine justice weighs the sins of the cold-blooded and the sins of the warm-hearted in different scales." That may be true, but purgatory is still the best either can hope for. In any event, neither Roosevelt's nor future liberal administrations would ever run short of warmheartedness. "Better the occasional faults of a government that lives in a spirit of charity than the consistent omissions of a government frozen in the ice of its own indifference," he thundered. But what if the faults of charitable government proved systemic rather than occasional? And besides, how charitable is it to fund warmheartedness with compulsory tax payments? This is liberality in Machiavelli's sense, being generous with other people's money.

Nonetheless, FDR saw as a *liberal* possibility the cultured, positive freedom that Wilson associated with well-rounded individuality in the State. Roosevelt had sloughed off all Germanic pretensions, of course, assuming he ever had any. He continued Wilson's own effort to make Progressivism breathe American air, and after World War I this was the impulse of most reformers. Not being an academic, FDR had much less to unlearn. In moving from German to English thinking, Protestantism was a convenient way station, and Wilson, whose father was a Presbyterian minister and his mother the daughter of a Presbyterian minister, fluently incorporated religious language and sentiments into his Progressivism. That was the era of the Social Gospel movement, a tributary of Progressivism, so it was common to encounter millenarian religious longings translated into calls for social work and social justice.[21] By the 1930s, however, this secularized Protestantism had been joined by more than one evangelical and fundamentalist resurgence, and when Roosevelt, as sensitive a barometer of his times as could be imagined, expressed the higher ethical life to which liberalism pointed, he did so in relatively unassuming, vaguely Protestant and vaguely Progressive terms that could appeal to almost everyone. He scourged the "money changers" in the temple of democracy, for example. "They know only the rules of a generation of self-seekers. They have no vision, and when there is no vision the people perish." "Happiness lies not in the mere possession of money," he intoned; "it lies in the joy of achievement, in the thrill of creative effort." America had come through "a period of loose thinking, descending morals, an era of selfishness," among government and people alike.

Many "amongst us have made obeisance to Mammon," permitting "the profits of speculation, the easy road without toil," to lure us from "the old barricades." To recover the higher standards of old "we must abandon the false prophets and seek new leaders of our own choosing." But those leaders will lead the country forward, not backward. "By using the new materials of social justice we have undertaken to erect on the old foundations a more enduring structure for the better use of future generations . . . ," he said. "We are beginning to wipe out the line that divides the practical from the ideal; and in so doing we are fashioning an instrument of unimagined power for the establishment of a morally better world."[22]

That the president should be in charge of leading Americans to a morally better world was not exactly a part of the office's original job description, which was to "faithfully execute the Office of President of the United States" and "preserve, protect, and defend the Constitution of the United States." Wilson had rewritten the oath, in effect, to make the president the leader of national public opinion—always assuming he was leading it into the future—rather than being merely an office-holding servant of the Constitution; and had subtly displaced loyalty to the written Constitution with loyalty to the unwritten, forever evolving one. Roosevelt was the first chief executive to have sufficient time and popular approval to put the new job description—the reformed duties and powers— into full effect. His unabashed partisanship functioned as a necessary means to the new ends, inasmuch as no president could lead the nation without having a means to galvanize and organize it for the march ahead. Both Wilson and FDR called this process of summoning and interpreting popular sentiment "common counsel," and it was possible only through the mechanism of the political party—a new kind of political party, to be sure, organized around the charisma or "personal force" of the leader and with as few intermediaries as possible between him and the people. Over time the bias was for a more ideological (or parliamentary-style) party system—liberals versus conservatives—just as FDR had suggested, because it is the logic of the liberal argument to divide politics between the party of the future and the party of the past, and because masses of people are more easily organized along such simple lines. The old statesmanship had always been a little embarrassed by partisanship—think of Washington's Farewell Address, or the custom of "standing" rather than "running" for office—but the new doctrines of leadership were very comfortable with it, so long as its overall effect was progressive and liberal. If politics were capable of achieving much more than George Washington could ever dream, after all, then the excesses of partisan zeal could be justified as the price to be paid for Transformation. As simultaneously a national and a party leader, because he couldn't be the former without being the latter, the president spoke for the people against the special interests, for the future against the past, and therefore for a morally better world as opposed to the same old world of imperfect human nature. Once Americans had crossed the threshold of the Promised Land, however, they presumably wouldn't need such partisan leadership anymore. To some extent, the New Deal imagined itself part of a final, vitriolic, partisan exchange that would usher in a postpartisan world dominated by enlightened administration and the welfare state.[23]

### New Deal and New Rights

FDR apparently didn't plan on the New Deal as the name of his program, but he embraced the moniker when the press did, seizing on a phrase from his 1932 acceptance speech: "I pledge you, I pledge myself, to a new deal for the American people." What was the New Deal? The term itself had an ambiguity that a politician could love. It could mean a new shuffle of the deck and a new deal of cards, the game and the rules remaining unchanged; or it could denote a wholly new arrangement, a partnership on different terms, a new social compact, as President Obama puts it. Roosevelt played the ambiguity brilliantly. In the end, the truth was closer to the second sense. The New Deal was about a new account of rights and a new account of the social contract, to which the old rules of the old game had to be accommodated.

With the New Deal, liberalism not only turned its attention to managing the economy and building the national welfare state, it rethought its methods and how its means related to its ends. Consider it in comparison to its predecessor reform period. The New Freedom had had no program or set of programs with which it was essentially identified. Wilson's first term had seen him master Congress in a virtually unprecedented way and win nearly every item on his domestic agenda: a significant lowering of tariffs; passage of the Federal Reserve Act, which created the "Fed"; the Clayton Act, which fleshed out the Sherman Antitrust Act; establishment of the Federal Trade Commission; and other laws that granted the eight-hour day to railway workers engaged in interstate commerce, put the government in the business of making farm loans, and more. All estimable reforms if you were a Progressive, but nothing to set the world on fire. His nimble turn to a more TR-style regulation of business as the 1916 election approached showed how overblown the family differences between the two candidates had become in 1912—like candidate Obama's opposition to the individual health care mandate in 2008 versus Hillary Clinton's support of it. But it would be hard to point to the Federal Reserve or the FTC as Wilson's crowning achievement or the New Freedom's essential legacy. One of the peculiarities of the living constitution was it seemed to imply the impermanency of all institutions and policies established under it. Save for the principle itself—the freest possible adjustment to new conditions through the coordinated action of the organs of government—everything else about the living constitution seemed in flux. That's the deepest reason that Wilson's programs seemed less important than his justification for them. What distinguished the New Freedom was the attitude of reform, the embrace of leadership and the living constitution, the anticipation of good things to come from opening American government up to progress, the idealism of expectation and the expectation of idealism. Its ultimate goal of fully satisfied and perfected individuality and sociality, of perfect social justice in the organic State, was a long way off, to say the least. Wilson's main reelection slogan in 1916 had nothing to do with his domestic reforms. It stated, *he kept us out of war.* When the war came, he invested in it all the New Freedom's idealism. "There is not a single selfish element, so far as I can see, in the cause we are fighting for," he said, as absurd a claim as could be made, for nations do not go to war without some interest of some kind being at stake, not even America. Wasn't making the world safe for democracy *in our interest*? America, he said during the debate on the Treaty of Versailles, "is the only idealistic Nation in the world."[24] By the war's end, the effect had been to discharge all that idealism like lightning down a lightning rod. It was gone, at least until the next big storm.

FDR had the advantage of Wilson's experience. He had seen numerous examples of "what an ill-advised shift from liberal to conservative leadership can do to an incompleted liberal program." After Taft succeeded TR, "little was left of the progress that had been made." "Think of the great liberal achievements of Woodrow Wilson's New Freedom," FDR said in a radio address in 1938, "and how quickly they were liquidated under President Harding. We have to have reasonable continuity in liberal government in order to get permanent results." Far more sinister than conservative presidents as a threat to permanent liberalism, however, was a growing economy. Already by his Second Inaugural FDR was worried that the economic rebound from the depths of the Depression in 1932 could spell trouble for the New Deal. Stagnation, despair, fear, and suffering had been propitious for reform; those "times were on the side of progress."

To hold to progress today, however, is more difficult.
Dulled conscience, irresponsibility, and ruthless self-
interest already reappear. Such symptoms of prosperity may
become portents of disaster! Prosperity already tests the
persistence of our progressive purpose.[25]

Impressed by the ebb and flow of liberalism, and keenly interested in "permanent results," Roosevelt decided to devise programs
and new institutions that would themselves bid fair to be permanent. These programmatic imperatives would be designed to fulfill
or secure new kinds of rights, socioeconomic rights, also meant to
be permanent, which he would enshrine in the so-called Second
Bill of Rights, a tip of the hat to the permanent status of the first
Bill of Rights. The new rights and their attendant programs would
come to constitute what he called "an economic constitutional
order," essential to securing economic democracy and social justice
for the long haul. As over against the formlessness of the living
constitution, FDR proposed a new kind of constitutional form
to secure these new socioeconomic or welfare rights. He moved
liberalism toward a two-tiered arrangement in which the written
Constitution remained subject to great informal revision in order
to accommodate the demands of progress; while the second, informal, economic constitution, though blessed with great inherent
flexibility, was built around unrepealable new rights and politically
inviolable benefit programs. Was this a case of wanting to have
his living constitution and eat it, too? Certainly it was a highly
original adaptation of the Progressive theory to liberal ends. Part of
its novelty came from the proud rehabilitation of individual rights
as part of the progressive project, effectively lifting both the first
and second bills of rights over the original text of the Constitution,
which contained all the conservative parts of the document that
the Progressives disliked (separation of powers, legislative checks

and balances, etc.). Broadly speaking, the Progressives had shunned rights talk because it reminded them of the old individualism and of the courts, each a big part of the problem as they saw it. But Rooseveltian liberals set out to recapture and renovate some aspects of old-fashioned constitutionalism for their own purposes, complicating but not abandoning the living constitution in doing so. In the long run, the tension between accommodating and encouraging political change as a general rule, and resisting every change—except expansions, of course—to core programs of the welfare state, would grow to bedevil liberalism.[26]

Herbert Croly in *The Promise of American Life*, published in 1909, had laboriously criticized the limits and contradictions of the American political tradition, and famously offered an alternative, or rather a synthesis, which his hero Teddy Roosevelt famously praised. Croly's formula was Hamiltonian means for Jeffersonian ends, that is, strong and flexible national government for the sake of human equality and brotherhood. The elementary point about a quasi-Hegelian dialectic like this was that the synthesis, which nullified, incorporated, and transcended the thesis and antithesis, was something altogether different from either of them. It was not a new shuffle of the same deck, a new combination of the old ingredients. The outcome was something that neither Hamilton nor Jefferson would have recognized, nor in all likelihood have approved. Croly understood and relished that point. So if TR's fifth cousin was following Croly's logic, as some historians say, he was not so much returning to the founding as he was leapfrogging over it. But FDR never said he was under Croly's spell, and his own statements deviated from Croly's line, treating the contest between Hamilton and Jefferson as still very much alive. As a political matter, FDR effusively embraced Jefferson, the patron saint of the Democratic Party, and shunned Hamilton as a master of public finance but a hopeless aristocrat. In 1925, in apparently the only book review he ever published, Roosevelt concluded his account of Claude

Bowers's *Jefferson and Hamilton* with the question: "Hamiltons we have today. Is a Jefferson on the horizon?" He was already casting himself in the role. When he returned to the question in his address to the Commonwealth Club, he depicted the two antagonists as generating a kind of cycle in American history so far, periods of Hamiltonian politics alternating with periods of Jeffersonian. Far from announcing a synthesis, he seemed inclined to agnosticism on the question whether Hamilton was right that individuals should "serve some system of government or economics," or Jefferson was correct that "a system of government and economics exists to serve" individuals. Honest men have differed, said FDR, "and for time immemorial it is probable" they will continue to differ on the issue. "The final word," he noted, "belongs to no man; yet we can still believe in change and in progress." As for himself he took Jefferson's side, or more precisely the Jeffersonian side updated; and as president, he had Jefferson's image put on the nickel, and he laid the cornerstone for and dedicated the Jefferson Memorial. No one was going to mistake Franklin Roosevelt, with his very public obeisances to Jefferson and the Declaration of Independence, for any kind of a Hamiltonian.[27]

Still, Mr. Jefferson had to be brought up to date, which Roosevelt proceeded to do, as he did many things, slyly. Rather than attack the notion of the social contract, as Wilson had done, he rehabilitated it. "The Declaration of Independence discusses the problem of Government in terms of a contract," he said to the Commonwealth Club.

> Government is a relation of give and take, a contract, perforce, if we would follow the thinking out of which it grew. Under such a contract rulers were accorded power, and the people consented to that power on consideration that they be accorded certain rights. The task of statesmanship has always been the re-definition of these rights in

terms of a changing and growing social order. New conditions impose new requirements upon Government and those who conduct Government.

The revisionism is quietly breathtaking. Government is presented as a contract between "rulers" and the people. In the Declaration's theory, there is no contract between the rulers and the people because neither exists ab initio; the whole concern is to make a people who can then choose a form of government that in their judgment will best "effect their Safety and Happiness." This government will represent them, not rule them. Gone from FDR's reading are the contracting parties of Jefferson's Declaration: the *individuals* who are "created equal" and "endowed by their Creator with certain unalienable rights." Gone too are the natural or God-given rights, possessed by those individuals. Missing as well is the people's explicit choice of a form of government. FDR did not bother to specify that the people may choose their rulers, or that the people may cashier their rulers if they violate individual rights and the people's safety and happiness. (At best these are inferences, weakened by FDR's description, earlier in the speech, of the slow process of political change: "The people sought a balancing—a limiting force. There came gradually, through town councils, trade guilds, national parliaments, by constitution and by popular participation and control, limitations on arbitrary power.") Original consent to government and the right of revolution are both silently passed over. In short, what Roosevelt presents here is an unattributed version of Wilson's account of political evolution or the development of constitutional government: no natural individuals, no natural rights, no original social contract, no right of revolution, but a gradual evolutionary process from autocracy to democracy, from no individual rights to a full panoply of them, mirroring the historical unfolding of society itself. The repudiation of Jeffersonianism could hardly have been more complete.[28]

FDR's new social contract began with rulers and people already in place, as though they existed from time immemorial. Under its terms, "rulers were accorded power, and the people consented to that power on consideration that they be accorded certain rights." The people as a collective have their rights as a result of the contract, the "give and take" with the rulers; individuals would hold their rights as members of the collective. But these rights don't precede society or the contract with the rulers, as they do in the Declaration; they come out of the contract. Popular rights and individual rights are an outcome of the negotiations; they are positive or prescriptive rights merely. Political liberty is then almost purely procedural, the right of the people to strike the balance that suits them between the power of the government and the privileges of the individual. As you might suspect, the model for this sort of social contract isn't the Declaration of Independence but Magna Carta, or something very much like Magna Carta.[29] Roosevelt didn't mention the Great Charter, however, because it would have complicated the historical analogy he wanted to draw, which was based on continental European not English history—namely, that disciplining the power of corporations today was like disciplining the power of unruly barons in feudal or early modern Europe. Both were tasks for a central government, which had to be strengthened precisely to protect the people's rights. The political lesson FDR sought to teach was that when the central government is on the side of the people, then according it more power will not diminish but enhance the people's rights. One could call this the First Law of Big Government: the more power we give the government, the more rights it will give us.

Yet what guaranteed that the central government will evermore be on the people's side? Roosevelt worried about periods of reaction, as we observed, and as the 1930s wore on he worried about the worldwide trend for nominal democracies to exchange their elections and liberties for fascist guarantees of prosperity and national greatness. But he seems never to have doubted that in the long run change and progress would win out, and that history guaranteed a better, and democratic, world. Democracy alone, he pointed out in his Third Inaugural Address, "has constructed an unlimited civilization capable of infinite progress in the improvement of human life." In his Fourth Inaugural he quoted his "old schoolmaster, Dr. Peabody," to similar effect: " 'The great fact to remember is that the trend of civilization itself is forever upward; that a line drawn through the middle of the peaks and valleys of the centuries always has an upward trend.' " For the very reason that democracy was the end product of centuries of slow evolution, the *finis ultimus* of all mankind's political development, its advance was reliable, and most reliable in those countries, such as the United States and Great Britain, where it had the deepest roots. When FDR said "political tyranny was wiped out at Philadelphia on July 4, 1776," he meant that the era of popular government was here to stay, and that the people could be trusted to administer their own government in their own interest. Tyranny of the majority was no longer the Achilles' heel of republican government. In fact, the pressing necessity was to stimulate the majority to take vigorous command of their government from the Tories, moneychangers, and "the dictatorship of the privileged few." "My program," he said in his 1932 acceptance speech, ". . . is based upon this simple moral principle: the welfare and the soundness of a nation depend first upon what the great mass of the people wish and need; and second, whether or not they are getting it." When he pledged his party to "the greatest good to the greatest number of our citizens," he indicated implicitly the utilitarian nature of the new power granted to government, and of the rights granted by government to the people, that is to say, to the majority, under his new version of the social contract.

As a "relation of give and take," government is a continual negotiation between the rulers and the people. As such, it will be one thing in one age and another in another, depending on what balance between governmental power and popular right is apposite. "The task of statesmanship," as we noted, "has always been the redefinition of these rights in terms of a changing and growing social order." For the first time, the liberal statesman's primary or characteristic task became to legislate a new morality, a novel vision of rights. Much depended, of course, on what it meant to "redefine" rights, and second, what it meant to redefine them "in terms of a changing and growing social order." The latter idea helped to reconcile the new social contract with the living constitution, through the medium of incremental adaptation—assuming that the redefinition was relatively minor. Less modestly, the formula reconciled the Hamiltonian and Jeffersonian ways of looking at the individual and the system discussed earlier in the Commonwealth Club address: under certain conditions, the individual may have to serve the government and the economic system; under others, the system will serve the individual. It depends on who must do the giving, and who gets to do the taking. In neither case, however, was the result undemocratic, in at least one sense of the term. Roosevelt went out of his way to assert that the people approved of the heavy hand of the European central government when it put down the haughty barons, even as they approved of the ruthless methods of the "financial Titans" who "pushed the railroads to the Pacific." The people could recognize the necessities of both situations, the utilitarian case for putting ordinary morality and sentimental populism aside to approve cruel action by the central government and by the private sector alike. History demanded no less. The people's own calculating pragmatism helped to excuse the sins of the monarchs and the millionaires. Eventually the kings or the financial Titans went too far, however, and outlived their usefulness. Then the social contract had to be renegotiated in a patently Jeffersonian way.[30]

In the American case, the president was perforce the statesman chiefly involved in redefining the people's rights. As party leader and national leader, the president's vision, Roosevelt implied, is not merely a general insight into what lies waiting to be developed or into the next step in political progress, but a specific insight into the new dispensation of rights demanded by "a changing and growing social order." When the people consented to governmental power in exchange for rights, they acknowledged and legitimized a new kind of unwritten presidential prerogative. How this prerogative power upset checks and balances or infringed on the separation of powers—for the power to redefine rights must be quasi-legislative, at least—was not a question for the people's or anyone else's deliberation because, like Wilson, FDR looked at government as a matter of *power* in the singular, not "powers" in the plural.[31] Rather than weighing its constitutionality or fidelity to republican principle, the people will judge the president's vision mainly by its effects, or in any case by its likely effects—just as decades before they had judged the financial Titans' building of the railroads and of America's industrial plant and pronounced it good, despite the greed and exploitation involved.

The continent-spanning railroads and mass-producing factories had been part of "a new dream" in the mid-nineteenth century. That dream "was the dream of an economic machine, able to raise the standard of living for everyone; to bring luxury within the reach of the humblest; . . . and to release everyone from the drudgery of the heaviest manual toil." In reality, the New Deal's socioeconomic rights will fulfill that dream, as the financial Titans never could. The people's new dream had arisen in response to the Industrial Revolution; the New Deal was launched in response to the changed society produced by that revolution and its crisis in the Great Depression. In neither case did the government or the people have much of a choice, he emphasized. "New conditions impose new requirements upon Government and those who conduct Government." Roosevelt wanted the New Deal's policies, including the redefinition of rights, to be seen as a necessary response to the Depression, not as his own choice or as his conclusions from certain ideological premises. But then it is always in the interest of willful executives to disguise their choices as necessities, to eschew "abstract" or "formalistic" arguments about morality and constitutionality in favor of the alleged dictates of the age. And besides, what choice did a leader have but to follow a people's inveterate habit of getting what they needed or desired, regardless of the formalities?[32]

Roosevelt's suggestion that Americans habitually had bent the rules of common morality shed a different light on his reassurance that "the terms of that contract are as old as the Republic, and as new as the new economic order." But such pragmatic compromises did not betray the higher morality because in the long run morality and economic self-interest converged. "We have always known that heedless self-interest was bad morals; we know now that it is bad economics," he said in the Second Inaugural. He held that the new rights, properly implemented, would prevent a future depression and stabilize democracy, in addition to improving American morals. He liked to quote "an old English judge" who said, " 'Necessitous men are not free men.' " Freedom required not merely the opportunity to make a living, FDR argued, but to make "a living decent according to the standard of the time, a living which gives man not only enough to live by, but something to live for." The old rights that supported (mere) political equality now seemed "meaningless in the face of economic inequality." Given the wealthy few's despotic control, "for too many of us life was no longer free; liberty no longer real; men could no longer follow the pursuit of happiness." To defeat this economic tyranny, Americans had to redefine the rights explicit and implicit in Jefferson's Declaration so as to establish economic democracy. The right to life must now be extended to include the "right to make a comfortable living," said Roosevelt. The right to property must be so revised as to guarantee the safety of men's savings and their well-being during childhood, sickness, and old age. Finally, liberty and the pursuit of happiness,

taken together, had to be reinterpreted to mean that the Jeffersonian "'rights of personal competency'—the right to read, to think,
to speak, to choose and live a mode of life, must be respected at all
hazards." No social compact, declared FDR, could justify depriv-
ing any individual of these rights so long as he respected the freedom of others.[33]

In his recasting of the Declaration, Roosevelt quietly severed
property, on the one hand, from liberty and life, on the other, dash-
ing the connection that John Locke and the American Founders
had labored to establish. For them, including Jefferson, the accumulation of property was a natural result of the use of an individual's right to life and liberty; and property functioned as an
early warning system against threats to liberty and life. Hamilton
and Jefferson agreed that property rights could be perverted and
turned against life and liberty—as in the great evil of slavery—
and that a maldistribution of wealth could result from aristocracy
and monopoly, which deprived individuals of their equal right to
pursue life, liberty, and property.[34] But neither Founder would have
accepted FDR's confidence that popular government had ceased
being a potential source of tyranny, that the accumulation of governmental power had ceased being a potential threat to individuals,
and that government (not nature or God) could be trusted to define,
much less redefine, men's rights. For Jefferson and Hamilton, the
"rights of personal competency" were the most natural rights of
all, being derived directly from the operations of the human mind.
Nature therefore remained a standard for these freedoms, and
reason her emissary; "the right to choose and live a mode of life"
had substantive and not merely formal restrictions. To vary the
point, for the Founders and Lincoln it wasn't merely the "complete
self-development" of "individual forces," as Wilson expressed it,
that defined happiness. One would have to speak of nature, rights,
God, talents, and virtues instead; of self-government, not merely
self-development; and of inequality as well as equality.

Roosevelt elaborated the new rights in his 1944 Annual Message to Congress proclaiming the Second Bill of Rights. World War II was still raging, but he had already begun to lay plans for the postwar establishment of "an American standard of living higher than ever before known." No matter how high that standard was, he maintained, if *any* fraction of the American people, even one-tenth, was "ill-fed, ill-clothed, ill-housed, and insecure," it could not be tolerated. The war against the Axis powers would soon be superseded by a war against poverty, by the return of Dr. New Deal.[35] He again resorted to the Declaration as the framework for the new rights. From its beginning, America had enjoyed "the protection of certain inalienable political rights"—free speech, free press, trial by jury, and other civil rights protected in the Bill of Rights—which constituted "our rights to life and liberty," he said. But as society changed, "these political rights proved inadequate to assure us equality in the pursuit of happiness. We have come to a clear realization of the fact that true individual freedom cannot exist without economic security and independence." To begin with, then, he turned what in the Declaration are the unalienable natural rights to life and liberty into the list of "political" or civil rights mentioned in the first Bill of Rights, though he deliberately excluded the Fifth Amendment's rights not to be deprived of one's property without due process of law, and not to have private property taken for public use without just compensation. Most of the rights he singled out were common-law rights given constitutional status and protection through the amendment process. He called them "inalienable," meaning not able to be taken away, though positive law (even the Constitution) can of course be changed, unlike the basic precepts of natural law. And he said that "this Republic had its beginning . . . under the protection" of these rights, though of course the Bill of Rights was not added to the Constitution until 1791. What he seemed to mean is that these were "inalienable political rights" in the sense of being part of immemorial prescriptive or common law, translated from British to American circumstances, and constituting the historical truth of the natural rights actually invoked in the Declaration. Roosevelt's new socio-economic rights were thus ripe fruits of the same historical tree, different from the original rights only in being up-to-date. "In our day," he said very precisely, "these economic truths have become accepted as self-evident. We have accepted, so to speak, a second Bill of Rights under which a new basis of security and prosperity can be established for all—regardless of station, race, or creed." To "become accepted" as self-evident is, for FDR's purposes, to *be* self-evident; self-evident is a historical category, not a logical or metaphysical one.

The "so to speak" Second Bill of Rights did not have to be argued for or even formally added to the Constitution. It was already "accepted," and as self-evident to boot. It was a kind of general historical revelation, obvious to anyone with experience of the times. "Among these" rights were the eight he propounded, which would help to structure liberalism's agenda for the next half century and more.

The right to a useful and remunerative job in the industries or shops or farms or mines of the nation;

The right to earn enough to provide adequate food and clothing and recreation;

The right of every farmer to raise and sell his products at a return which will give him and his family a decent living;

The right of every businessman, large and small, to trade in an atmosphere of freedom from unfair competition and domination by monopolies at home or abroad;

The right of every family to a decent home;

The right to adequate medical care and the opportunity to achieve and enjoy good health;

The right to adequate protection from the economic
fears of old age, sickness, accident, and unemployment;
The right to a good education.

"All of these rights spell security," he concluded, and when implemented will advance the nation toward "new goals of human
happiness and well-being." Security was the key, because "without
economic security and independence" individual freedom cannot
exist. In fact, the self-evidence or obviousness of these rights was
not obvious at all, and they badly needed a better argument than
Roosevelt—one cannot say gave, but rather alluded to. To begin
with, the rights conflicted with one another. The farmers' right to
sell their crops and livestock at a price high enough to provide a
decent living implied, and the New Deal agricultural programs put
into effect, controls on the amount of production. Constrict the
supply of food, and the prices farmers receive will rise—but so will
the prices consumers pay, eroding the "right to earn enough to provide" adequate food, clothing, and recreation. And what was the
point of lowering food production in a country one-third of whose
citizens were "ill-fed"? Similarly, the right to adequate earnings led
to policies like the federal minimum wage, which made it more expensive to hire new workers in an era of extremely high unemployment, despite the right to a job.[36]

Second, the status of these "rights" qua rights was mysterious.
Undoubtedly, the objects demanded in the Second Bill of Rights
were worthy—useful and remunerative jobs, good earnings, decent
homes, adequate medical care, good education, and so forth—and
many of them could be advanced by sound local, state, or federal
government policies, even if primary responsibility for the provision of these goods remained in private hands. That would have
been the traditional way of looking at them, measuring each good
against available resources, other claims on those resources, the
Constitution's stipulations on what government could and could

not do, the possible unintended consequences of policies encouraging them, and the policies' likely effects on the economy and on citizens' character. An ample supply of goods like these was conducive to a thriving middle class, and a thriving middle class was conducive to democracy. Statesmen had understood as much since Aristotle's *Politics*. But it didn't necessarily follow that the best way to furnish these pleasant, useful, and occasionally noble things was through a planned economy and government ukases. But Roosevelt insisted on turning these goods, whose attainment and use would have had to be governed by prudence, into rights, another animal entirely. Rights are the reciprocal of duties. Who exactly had the corresponding *duties* to supply these jobs, living wages, decent homes, and adequate medical care? The immediate answer was Congress. It is "definitely the responsibility of Congress" to implement "this economic Bill of Rights," Roosevelt declared. Proposed legislation was in some cases already moving through congressional committees. The legislature had to turn these rights into programs, which the government, or rather society, or rather the part of society not receiving the boon, would presumably fund. Little consideration was given to casting these rights into actual amendments to the Constitution that supermajorities of the Congress and the state legislatures would have to pass. With a living constitution, those arduous legalities could be dispensed with. Supermajorities of Congress alone, acting in parliamentary style, were adequate to the task of informally changing the country's formal, and fundamental, law. Besides, in order to be implemented the new rights would require legislation and hundreds (eventually thousands) of pages of regulations. These regulations and laws increasingly served as the nation's effective constitution.

Although not all of the new rights have been written into law, Social Security and other entitlement programs emerged eventually to guarantee adequate food, clothing, protection from the economic fears of old age, sickness, unemployment, and so forth. Because they discharged government's or society's duty to rights holders, entitlement programs' spending could not be subject to the normal budgetary process, which pitted program against program, seeking to weigh each against other claims on spending and ultimately against the common good. But "rights" could not be weighed, balanced, or deliberated. Entitlement programs had a moral imperative all their own and soon became, as FDR had foreseen and desired, a kind of government within the government. *Welfare state*, a term he didn't use, is nevertheless an apt description of the new kind of state that emerged within or alongside the old government under the Constitution and the original Bill of Rights. On the budgetary front, an even more ominous term had to be invented to describe the automatic increases authorized by statute law to fulfill government's duty to its needy and importunate citizens: "uncontrollable spending." It's hard to see how the Republic can be self-governing if every year a majority, a growing majority, of the federal budget consists of uncontrollable spending. But then one doesn't ask how much freedom of speech is costing this year, so why worry about Social Security?

The answer goes to the difference between the old rights and the new. Sometimes the difference is said to be between negative and positive freedom. Man's natural rights, and the civil rights based on them, typically culminate in Thou-shalt-nots: do not murder, because it violates the right to life; do not steal, because it violates the right to property; and so forth. Positive rights embody not freedom *from* evils but freedom *for* various good things, rights to have something good or do something good: Thou shalt have a decent home, and thy neighbor shall provide it. There's some utility in the distinction between the two kinds of rights, but it breaks down eventually.[37] For example, the Sixth Amendment's right to "a speedy and public trial" in criminal cases is not a negative but a positive right, even though it could be said to be necessary to protect your negative right to be free from unjust punishment. But that shows just how entwined the two kinds of freedom are. It's more accurate to take the old and new rights at face value. The old rights in the Declaration were unalienable because they were natural rights, aspects of the kind of beings that humans are by nature, neither God nor beasts, rational but with a full complement of passions and needs. Human nature was partly fixed—particularly in its rights—and partly not, capable of using its endowment of rights to form different customs and laws. Men's equality made them, as adults, equally free of one another's authority, and so possessed of equal rights to life, liberty, and the pursuit of happiness. The state of nature, derided by Woodrow Wilson as a historical nullity or impossibility, was a hypothesis designed to exhibit man's nature absent custom and law, and to explain why and how man would convert insecure natural rights and liberty, by an act of enlightened consent, into the peace and security of civil rights and liberty. Guided by interest and right, citizens were responsible for their own use of their liberties, and the economic and social inequalities that would inevitably result. Still, these rights implied correlative duties to other human beings, and to the country as a whole.[38]

Roosevelt's rights were not "natural"—he abandoned the distinction between natural and civil rights, for instance—but historical, complements of liberating social change and evolution, and thus better called "human rights" because they had been made by man himself over time and had become necessary to the full humanity of modern men in progressive societies. Like Teddy Roosevelt and Wilson before him, FDR liked to oppose "human rights" to "property rights."[39] The latter he regarded as dangerous abstractions, frozen in time and icily indifferent to actual human needs and development. It isn't much of an exaggeration to say that, from the Progressive point of view, the quintessential property right was the right of property in slaves—the quintessential *violation* of property rights, in the contrary view of Lincoln and Jefferson. To liberal statesmen and historians, the Civil War was the victory of human rights over property rights—not a victory for man's natural rights, including self-ownership and the other rights to hold and acquire property. Though he never condemned private property as such, and indeed was eager to see decent homes, prosperous farms, and remunerative jobs in the hands of the people, Roosevelt was clear that the new rights had to be paid for by *someone*, and he was eager to compel the rich to do it. His explanation, both moral and economic, of the Depression traced it to a maldistribution of income and thus of purchasing power between the rich and the poor. The problem was "underconsumption," as he called it. During the 1920s, the rich had gotten richer and the poor poorer; by 1929 there were too few buyers for too many goods, and the unhappy result was the Depression. The rich could afford things, to be sure, but how many automobiles and iceboxes did a plutocrat need? Following the Brains Trust's advice, FDR prescribed "adjusting production to consumption," which meant national planning on an immense scale, and "distributing wealth and products more equitably," which meant income redistribution on an unknown scale. "We have no actual famine or dearth," he explained; "our industrial and agricultural mechanism can produce enough and to spare." No supply-side solutions were necessary; the malady was entirely on the demand side, and the cure, too. "Our government formal and informal, political and economic, owes to everyone an avenue to possess himself of a portion of that plenty sufficient for his needs, through his own work," if possible. The new rights were intended to hasten the transfer of "a portion of that plenty" from the rich to the poor, revitalizing the economy and American morality at the same time.[40]

As economic theory, underconsumption was underwhelming. The rich had gotten only slightly richer during the 1920s, corporate profits were steady (as a percentage of revenues), employee wages actually had risen from 55 to 60 percent of corporate income during that decade, and the percentage of GNP going to consumption expenses had risen, too, from 68 to 75 percent. "The ratio of consumption to national income was not falling in the 1920s," bluntly concluded Peter Temin, an economist at the Massachusetts Institute of Technology. "An underconsumptionist view of the 1920s, therefore, is untenable." In fact, "The concept of underconsumption has been abandoned in modern discussions of macroeconomics."[41] There was another uncomfortable fact: the rich were not rich enough to fund the welfare state on their own. They were few, as FDR kept emphasizing—"5,000 men in effect control American industry," he said on the campaign trail in Boston; "at the end of another century we shall have all American industry controlled by a dozen corporations, and run by perhaps a hundred men," he prophesied to the Commonwealth Club—but the middle class was legion and had the bulk of the income. Even with top marginal income tax rates on the very wealthy jacked up to 79 percent and the marginal estate tax on the rich at 70 percent, the Treasury did not generate enough revenue to fund the new rights. So the duty to pay for them ineluctably shifted to the middle class. For the most part, New Deal programs were funded by regressive excise taxes on cigarettes, gasoline, and alcohol (once Prohibition had been repealed), the regressive Social Security tax, and others that hit the poor and middle class hardest. The money floated out of the majority's right pocket and over into its left, minus the costs of the trip to and from Washington, D.C.[42]

Most historians describe the New Deal as a great experiment in pragmatic governance.[43] At Oglethorpe University in 1932, Roosevelt had declared, "The country needs and, unless I mistake its temper, the country demands bold, persistent experimentation. It is common sense to take a method and try it: If it fails, admit it frankly and try another. But above all, try something." A lot of the New Deal policies *were* improvisational. Especially in the first Hundred Days, it was hard to know where the wheels of the acronym slot machine were going to stop—WPA, PWA, AAA, NIA, etc. But the tenor and aims of the policies were not made up on the run. Two paragraphs before that famous remark, he said, "Let us not confuse objectives with methods." He was calling for experimentation in the means to economic security and prosperity, not in the New Deal's goals; and besides, as with most liberals who preach pragmatic experimentation, he didn't practice it all that readily.[44] The list of programs his administration tried and discontinued because they were failures, or for that matter successes, is short. Of the major programs shuttered during the New Deal, most were struck down by the Supreme Court as unconstitutional—most notoriously, the National Recovery Administration and the Agricultural Adjustment Administration. Their abolition redounded to FDR's long-term benefit. If they had lived, his later reputation as a moderate pragmatist would have suffered—as would the economy in the 1930s. Still, one should acknowledge the extent to which many even of the apparently improvised programs had been thought up, if not through, in advance. Never one to let an emergency go to waste, Roosevelt seized the opportunity to ram through plans and the outlines of plans that had been on the shelf sometimes since the Wilson administration, and that often had nothing directly to do with economic recovery. As the historian Daniel Rodgers explains:

> But to a striking degree the New Deal enlisted its ideas
> and agenda out of the progressive past. Old-age and unemployment insurance, public housing, the National Labor
> Relations Act, the Fair Labor Standards Act, emergency
> work relief, rural electrification, banking and securities
> regulation, holding company legislation, and agricultural reform all had precrisis roots. In style, in urgency,
> in federal-state relations, and in political alliances there
> was no missing the administration's new departures. But
> as a legislative program, it is far more accurate to see the
> New Deal as a culmination: a great gathering in from the

progressive political wings of a generation of proposals and
ideas.[45]

The Second Bill of Rights was in many ways an unpragmatic
measure. It was meant to be constitutionalized, albeit in a living
constitution capable of easy change; but the point of calling it a bill
of rights was to suggest its permanence and to associate it with the
veneration and "blind worship" that attached to the actual U.S.
Constitution.[46] Roosevelt defended it not only as useful to economic recovery but also as demanded by latter-day human decency.
Nonetheless, its rights promised streams of material benefits of various kinds, and not to everyone equally. Only a developed and prosperous society could afford them in the long or the short run. To
most of the world's countries, especially in the 1930s, its promises
would read as bitter satire. True to the Progressive understanding,
these "rights" had to vary depending on a nation's stage of development, its national spirit, its wealth; they were not universal. No
country was too poor to guarantee freedom of worship but only
a few were rich enough to enforce, or fancy they could enforce,
the right to adequate medical care. Moreover, nearly all the Bill's
benefits were targeted—to farmers, businessmen, families, the old,
the sick, the unemployed, and so on. Individuals became eligible
for economic rights if they fell into one of these groups. The indi-
rect but hardly unanticipated consequence was to encourage people
to think of themselves as members of such groups, and to organize politically to lobby for more rights, which is to say, streams of
benefits. Almost immediately the high idealism of socioeconomic
rights descended into the grubby politics of interest groups. Instead
of equal rights for all, Irving Kristol noted, the New Deal's slogan
became in effect "equal privileges for all." Every American had an
interest now in becoming part of a vested interest, a special interest.
But these groups' rights lacked "any corresponding set of obliga-
tions" to restrain or moralize them. The old individualism had a

moral discipline that ranged from the duty to respect others' rights to a sense of patriotism and civic virtue. By comparison, the new interest groups were, in Kristol's words, "more crass and vulgar and materialistic" as well as less sentimental. "After all, one knows sentimental businessmen, sentimental trade union leaders, sentimental professors even. But there is no such thing as a sentimental corporation, a sentimental trade union, or a sentimental university." With so few moral restraints and so many possible rights, interest-group liberalism found the only sort of balance it could, "the equilibrium of the jungle," Kristol called it, of countervailing power among the competing factions.[47]

"Selfishness is without doubt the greatest danger that confronts our beloved country today," Roosevelt once said.[48] He hoped that by liberating Americans from necessitousness—by satisfying basic human needs and desires through a planned and managed economy, and through the welfare state based on a new doctrine of economic rights—he could usher in a new liberal age, distinguished by less selfishness and materialism. And he succeeded to some extent in elevating American morale through the long depression he never cured. But rather than permanently lifting the moral tone of American life, the welfare and regulatory state plunged our politics into an amoral scramble for power, benefits, and influence that looked all the more tawdry next to the high hopes Roosevelt had raised. This interest-group or social-welfare Darwinism proved a lasting part of liberalism, and helped spur the greater disillusionment to come in the 1960s.

**4**

___

# Lyndon B. Johnson and the Politics of Meaning

**B**arack Obama is a child of the Sixties, born in 1961, the same year that the word *lifestyle* entered *Webster's Dictionary*. By most definitions, he's a baby boomer, though he hasn't dwelled on that status and sometimes seems to reject it. "In the back-and-forth between Clinton and Gingrich, and in the elections of 2000 and 2004," he writes in *The Audacity of Hope*, "I sometimes felt as if I were watching the psychodrama of the Baby Boom generation—a tale rooted in old grudges and revenge plots hatched on a handful of college campuses long ago—played out on the national stage." As a late boomer, Obama is much younger than Clinton, born in 1946, and Gingrich, born in 1943, and obviously didn't attend college in the 1960s, as they did. In addition, he has substantive reasons for wanting to distance himself from the "grudges and revenge plots" born in that decade. "The victories that the sixties generation brought about—the admission of minorities and women into full citizenship, the strengthening of individual liberties and the healthy willingness to question authority—have made America a far better place for all her citizens," he affirms. But "those shared assumptions . . . that bring us together as Americans" were lost in the process and have not been replaced.[1]

In one sense, as he admits, he's "a pure product of that era." As "the child of a mixed marriage," his opportunities would have been "entirely foreclosed" without the "social upheavals" of those years. At the time he was too young to understand that, but as an adolescent in the 1970s he connected to the Sixties firsthand through his own rebellion. "I became fascinated with the Dionysian, up-for-grabs quality of the era, and through books, film, and music," and eventually marijuana and cocaine, he reenacted the Sixties in his own life. *His* Sixties was not the Peace Corps and Freedom Rides, but "Huey Newton, the '68 Democratic National Convention, the Saigon airlift, and the Stones at Altamont." In college, he says, he began to see "how any challenge to convention harbored within it the possibility of its own excesses and its own orthodoxy," and so he reexamined his "assumptions" and discovered the pleasures of the Apollonian. His own experience left him, however, with a sense of the lasting difference the Sixties had made in American politics.

Despite a forty-year remove, the tumult of the sixties and the subsequent backlash continues to drive our political discourse. Partly it underscores how deeply felt the conflicts of the sixties must have been for the men and women who came of age at that time, and the degree to which the arguments of the era were understood not simply as political disputes but as individual choices that defined personal identity and moral standing.

I suppose it also highlights the fact that the flash-point issues of the sixties were never fully resolved. The fury of the counterculture may have dissipated into consumerism, lifestyle choices, and musical preferences rather than political commitments, but the problems of race, war, poverty, and relations between the sexes did not go away.[2]

He regards as odd the depth and divisiveness of Sixties politics, the way it involved "personal identity and moral standing," but that is the way with civil wars, and the Sixties was a double civil war, within liberalism itself and then between both kinds of liberalism and what Nixon would call the silent majority. For in that decade the radicalism that had been latent all along in liberalism broke free of its faith in progress, in science, and in the democratic process itself. Mainstream liberalism found it hard to confront the radicals because so many of their premises were its, too. The very familiarity of the New Left's arguments astonished and silenced Lyndon Johnson and many of his supporters.

Where Nixon had failed, Ronald Reagan succeeded, using the Sixties polarities to stamp liberalism as unpatriotic, and conservatism as the natural home of the patriotic majority. Obama admired Reagan's skill and studied his example closely, even as he deplored his policies. Bill Clinton, the first post-Reagan Democratic leader to try to transcend the ideological categories as Reagan had defined them, fell short because, as Obama astutely observes, Clinton's messy personal life "could be made to embody the very traits of sixties liberalism that had helped spur the conservative movement in the first place." Taking that lesson to heart, and learning from his own experience, Obama is at pains to be, and to be seen as, a strong family man, a responsible husband and father urging responsibility on others, a patriot, a model of pre-Sixties, subliminally anti-Sixties, sobriety. His demeanor and even his suits, though not his skin color, resemble something you'd see on *Father Knows Best*.[3] Allergic as he is to the social disorder and extremist politics of the Sixties Left, Obama nonetheless admires its daring and its results. He praises LBJ's historic breakthroughs in the Great Society, too, though reservedly due to Johnson's imperfect record on racial equality before the Civil Rights Act of 1964—and because his policies on the Vietnam War, among other things, precipitated the great liberal crack-up, the unraveling of the New Deal coalition.[4]

Obama tries to stand on both sides of the liberal civil war, and on neither. He wants to put the pieces back together again. He seeks to reunite, in Richard Rorty's terms, the academic-cultural Left and the reformist-political Left. It won't be easy, for one reason because he hasn't been able to reconcile them in his own mind.

## The Great Society

Campaigning in Providence, Rhode Island, in September 1964, Lyndon Johnson climbed onto the roof of his presidential limo and proclaimed, "And I just want to tell you this: we're in favor of a lot of things and we're against mighty few."[5] Welcome to the Great Society. Though she wasn't talking politics, Mae West caught the spirit of the enterprise: "Excess is not nearly enough." Johnson was not meant to be president—certainly not by John F. Kennedy or Bobby Kennedy—but after ascending to the office he decided quickly that President Kennedy's assassination had created an opportunity for sweeping political and social reform. He resolved to enact the languishing parts of the slain president's agenda (tax cuts, a modest antipoverty program, civil rights legislation) and use that as a "springboard" to "create a Johnson program, different in tone, fighting and aggressive. Hell, we've barely begun to solve our problems," he told Richard Goodwin, the former Kennedy aide who became his chief speechwriter, as they floated in the presidential swimming pool. "And we can do it all. We've got the wherewithal." In one of those infamous interviews during which Johnson was sitting on the presidential toilet, he repeated the sentiment. "Hell, we're the richest country in the world, the most powerful. We can do it all, if we're not too greedy. . . ." That was the agenda LBJ had in mind, to "do it all," solve every one of America's social problems. "These are the most hopeful times in all the years since Christ was born in Bethlehem," he said in December, having been elected by historic margins the preceding month. For "today—as never before—man has in his possession the capacities to end war and preserve peace, to eradicate poverty and share abundance, to over-come the diseases that have afflicted the human race and permit all mankind to enjoy their promise in life on this earth."[6]

Johnson's political career had begun under FDR's patronage, as the Texas director of the National Youth Administration, and then as a Congressman. "Franklin D. Roosevelt, he was my hero, he was like a father to me," Johnson told Walter Cronkite. Yet the son desired nothing so much as to surpass the father. LBJ ordered his aides to count the number of times his Inaugural Address was interrupted by applause to see if it exceeded the previous high set by Roosevelt in 1933 (it didn't). He raced to get more bills passed in the hundred days after inauguration than FDR had in his Hundred Days, and succeeded. In late 1966 he boasted, "FDR passed five major bills [during] the first one hundred days. We passed two hundred in the last two years. . . ." Goodwin knew his man: "He wanted to out-Roosevelt Roosevelt."[7]

In the first place, that meant completing the New Deal's agenda of what Arthur Schlesinger Jr. called *quantitative* liberalism, guar-anteeing economic security and opportunity through such new programs as Medicare and food stamps. To "finish that work" LBJ called for an unconditional "war on poverty," whose goal, he said characteristically, was "total victory." He emphasized that the idea wasn't to relieve poverty "but to cure it and, above all, prevent it"; a "lack of jobs and money" was but the "symptom" of poverty, not its cause. In the "affluent society" of postwar America, to use John Kenneth Galbraith's phrase, the poor were profoundly differ-ent from the tens of millions who had been plunged into prolonged, but still temporary, poverty by the Great Depression. The new poor were a separate nation, estranged and apart from the vast middle class. In the title of another fashionable book of the day, they were "the other America." The stubborn persistence of poverty amid un-precedented prosperity suggested, at least to midcentury liberals, that the poor were victims of a deep problem with society itself.

To rescue them, it would be necessary to change society as a whole. LBJ was eager to sign up for that larger goal, despite the advice of many of his advisors to stick to a modest expansion of New Deal programs. He turned down the entreaties of Jack Valenti, George Reedy, and others in favor of Goodwin's and Bill Moyers's recommendation for a transformative agenda, a cultural revolution. And so in his signature speech Johnson launched the first of many attacks on the nation's underlying condition, the deep-seated problem not confined to the poor and the old: the spiritual impoverishment of American society as a whole. "For in your time," he told the graduating class at the University of Michigan in May 1964, "we have the opportunity to move not only toward the rich society and the powerful society, but upward to the Great Society." Liberalism now had to turn its attention to improving "the quality of our American civilization." The age of *qualitative* liberalism had begun.[8]

To be sure, qualitative reform presupposed that quantitative goals had been or were being met. The Great Society presumed "abundance and liberty for all," and so demanded "an end to poverty and racial injustice," to which the Johnson administration was (naturally) "totally committed." (There was nothing halfhearted about this administration, except, that is, for its prosecution of the Vietnam War.) But that was "just the beginning." As Johnson explained:

> The Great Society is a place where every child can find knowledge to enrich his mind and to enlarge his talents. It is a place where leisure is a welcome chance to build and reflect, not a feared cause of boredom and restlessness. It is a place where the city of man serves not only the needs of the body and the demands of commerce but the desire for beauty and the hunger for community.
>
> It is a place where man can renew contact with nature. It is a place which honors creation for its own sake and for

what it adds to the understanding of the race. It is a place where men are more concerned with the quality of their goals than the quantity of their goods.

But most of all, the Great Society is not a safe harbor, a resting place, a final objective, a finished work. It is a challenge constantly renewed, beckoning us toward a destiny where the meaning of our lives matches the marvelous products of our labor.

Here then was LBJ's vision of the future. Ending poverty and racial injustice were mere appetizers. The main course was creating a society that satisfied every person's desire for learning, work, material security, creativity, contact with nature, beauty, community, and meaning. Leaving aside the jokes at the beginning, the speech quoted only two people, Woodrow Wilson and Aristotle. The former testified (he was president of Princeton University at the time he spoke these words) that "every person sent out from his university should be a man of his nation as well as a man of his times." Aristotle was cited from the *Politics*: "Men come together in cities in order to live, but they remain together in order to live the good life."[9]

But can you live the good life if you are merely a person of your nation and your times? What if your nation and times are bad? Like the Progressives, Johnson wanted to return to Aristotle's so-to-speak communitarian and idealist ethos in order to overcome America's commercialism and materialism. But the return was spoiled from the outset. Aristotle was neither a mere communitarian nor an idealist. According to him, the two highest human activities are politics and philosophy. There is no politics and no philosophy in LBJ's description of the "Great Society," no questions of who should rule and for what ends, no contest over beauty, community, and meaning, no standards by which to measure the "quality" of men's goals, no truthful "resting place" or "final objective." Johnson's assumption was that our never-ending quest for meaning will discover good and happy things because history itself guarantees good and happy experiences, once the oppressions of the past are cleared out of the way. More precisely, his words suggested that once necessity in the form of scarcity, poverty, and war has been conquered, we will be free to make or create the meaning of our own lives, thus matching "the marvelous products of our labor." Far from overcoming commercialism and materialism, the Great Society invited us, individually and collectively, to go into the lifestyle business—to make and remake ourselves in accordance with our own values and desires.

Collectively, this meant that the Great Society accepted and encouraged—how consciously is a different question—the New Left's inclination to base politics increasingly on issues of personal identity, gender, and sexuality, and "postmaterialist" concerns like environmental activism. New kinds of interest groups representing these causes would arise in the late 1960s and 1970s partly in response to this prompting, and often cutting against traditional Democratic interests like city machines and labor unions. The new environmentalism raised in particularly stark form the issue of what "progress" entailed and whether liberals were for it or agin' it: here the clash between quantitative and qualitative, blue-collar and white-collar, modern and postmodern agendas was manifest. Although such tensions helped to shatter the old Democratic coalition, and very quickly too, they nevertheless obeyed the logic of the old liberal State, with its open-ended sympathy with evolving social forces and with individual self-development.

It was the young who experimented with the new causes and lifestyles, not Johnson, and the incongruity of him quoting Aristotle and extolling the life of the mind and the desire for beauty was not lost on them, or on him. The very boldness of his speech, its emphasis on the qualitative enrichment of life, and the high-sounding goals it set for the Great Society, opened him to many bad and unhappy experiences later on. Above all, it made him increasingly dependent on the very people for whom he expressed the most contempt: the "Harvards." These were the liberal intellectuals, especially those associated with the Kennedys, who looked down their nose at the graduate of San Marcos State Teachers College in Texas. The "best and the brightest" had flocked to the Kennedy administration and Johnson craved their approval even as he returned their disdain. In the Great Society he thought he was giving them all they could have asked for, and much more than JFK would ever have managed; but he had little idea of how much they were prepared to ask for. To run the War on Poverty and the Great Society, and even to devise the programs that constituted them, he needed the Harvards' help. In fact, about the only practical action the Great Society speech recommended was "a series of White House conferences and meetings" to discuss what to do about the problems of the cities, natural beauty, education, and "other emerging challenges." At these sessions he promised "to assemble the best thought and the broadest knowledge from all over the world to find those answers for America." Social scientists had been a crucial inspiration for Progressivism, and had had through the Brains Trust and the agencies engendered by it a significant impact on public policy in the 1930s. Under President Eisenhower, the new Department of Health, Education, and Welfare had given them a permanent platform in government. But in the Great Society they came into their own. The Johnson administration created an unprecedented number of task forces, stocked with pretend as well as genuine experts, to tackle the nation's social problems. As Samuel Beer, an impressive liberal political scientist, explained, "the antipoverty program was not shaped by the demands of pressure groups and the poor—there were none—but by deliberations of [White House] task forces. . . ." "In the early 1960s in Washington," Daniel Patrick Moynihan, another impressive liberal, wrote, "we thought we could do anything. . . ." Social science encouraged and validated "the central psychological proposition of liberalism . . . that for every problem there is a solution." Unfortunately, the solutions in most instances made the problems worse. It took time for even as sensible a political scientist as Moynihan to realize, as he admitted later, "*The government did not know what it was doing.* It had a theory. Or rather, a set of theories. Nothing more."[10]

Having staked the success of his administration on those whom he distrusted and whose knowledge he routinely scoffed at—"kooks and sociologists" was one of his milder descriptions—he then staked his fortunes on yet another unsteady ally: the young. He challenged those University of Michigan graduates: "Your imagination, your initiative, and your indignation will determine whether we build a society where progress is the servant of our needs, or a society where old values and new visions are buried under unbridled growth." Your *indignation*—you can't say he didn't ask for it. When the campuses erupted in protests against the war, and the "days of rage" interrupted the Summer of Love, and the demonstrators in Grant Park clashed with police outside the Democratic National Convention—the president may have regretted his choice of words. "For better or for worse," he added—words that would echo ominously over his administration's future—"your generation has been appointed by history to deal with those problems and to lead America toward a new age. You have the chance never before afforded to any people in any age. You can help build a society where the demands of morality, and the needs of the spirit, can be realized in the life of the nation." He concluded the speech with four "will you join in the battle" questions to his young audience, who responded with "mounting applause" and "mingled shouts of affirmation," in Goodwin's account; "the mammoth Texan had, for the moment, transformed the worldly northern university into a Baptist meeting."[11]

Johnson expected the applause to continue because he and the kids were on the same page, condemning "soulless wealth" and seeking to realize a vision of "a new world" without poverty, racial injustice, war—"a way of life beyond the realm of our experience, almost beyond the bounds of our imagination." Join me, he told them in effect, in going boldly where no man has gone before. After he left office five years later and went into exile in Texas, he plaintively told Doris Kearns (now Doris Kearns Goodwin), "I just don't understand those young people. Don't they realize I'm really one of them? I always hated cops when I was a kid, and just like them I dropped out of school and took off for California. I'm not some conformist middle-class personality. I could never be bureaucratized."[12] Shortly after his announcement that he would not run for reelection, he had explained his reasons to Vice President Hubert Humphrey: "I could not be the rallying force to unite the country and meet the problems confronted by the nation abroad and at home in the face of a contentious campaign and the negative attitudes towards [me] of the youth, Negroes, and academics." The youth, Negroes, and academics—all the groups whose approbation he had desperately sought.[13]

Most analysts think, in his speechwriter Goodwin's words, that "Lyndon Johnson's war destroyed Lyndon Johnson's Great Society." Goodwin himself doubts this is the whole truth. The deepest truth, actually, is that the Great Society destroyed the Great Society. Its soaring expectations, its utopian promises, could not be fulfilled in ten years or a hundred years. What it proffered was the satisfaction, in principle, of all material and spiritual needs and desires. But human desires are infinite. They cannot be satisfied, unless they are first governed or moderated by reason and morality. Its youthful audience and often the administration's own experts expected paradise now, egged on by LBJ's own intoxicating words. "Having the power, we have the duty," as he put it succinctly in announcing the War on Poverty.[14] If the abolition of poverty was possible, then it was a moral necessity, and the persistence of poverty, even at low levels, was morally intolerable. Perversely, the lower the poverty rate or the unemployment rate, the more intolerable it seemed. The final obstacle to a perfect world was 5.2 percent unemployment? What kind of a sick society refused to solve that problem?

Within American liberalism, the hope that the State could, and would, bring about the complete spiritual fulfillment of its citizens was not unique to Johnson. It first bubbled up in the Hegelian insistence that only in the rational State could the complete ethical life (*Sittlichkeit*), combining subjective and objective right, be lived. Under the spreading influence of Alexander von Humboldt, John Stuart Mill, and T. H. Green, that doctrine began to focus more and more on the maximum development of every person's faculties and capabilities as the criterion of a free and progressive society. So in America in 1886 Richard Ely described the "ethical ideal" of political economy, indeed of sociology in general, as "the most perfect development of all human faculties in each individual, which can be attained." Comparing this objective to the Christian parable of the talents and Kant's categorical imperative for each rational being to develop his talents fully, Ely advocated "such a distribution of economic goods" as would nurture the "growth of all the higher faculties—faculties of love, of knowledge, of aesthetic perception, and the like, as exhibited in religion, art, language, literature, science, social and political life."[15] What "distribution of economic goods" was it, then, that would nurture, say, the religious life? Ely didn't say, because he didn't know, though he implied it would be a more egalitarian one. Experience suggests that the wealthy are not always the most pure-hearted, and that the poor are quite capable of living pious and decent lives. But the Progressive argument suggested otherwise. Only when "poverty men," as Ely's student Simon Patten called them, are lifted out of poverty will they "adjust themselves as capably as normal men and . . . also appreciate culture and morality."[16]

The right to complete self-development, as Woodrow Wilson called it, came to full fruition and indeed to overripeness in the

Great Society. The New Deal, as we've seen, had been determined to drive the moneychangers from the nation's temples, and it had paid attention as well to the spiritual hygiene of men who were no longer "necessitous." When the Works Progress Administration paid artists to paint murals in post offices around the country, for example, it was partly to help the artists but also partly to raise citizens' aesthetic sense. Even before his presidency, FDR had been interested in resettling city dwellers onto farms to improve their health and morals, and as president he experimented with "communities programs" that built a hundred or so small cities around the country, models of cooperative living, close to nature, stocked with music teachers and art instructors and books by John Dewey and other opportunities for the creative use of leisure. The New Dealers had "their Heavenly City," too, as historian William Leuchtenburg noted.[17]

But it was the Great Society that launched the third wave of liberal reform, squarely committed not only to the full satisfaction of material needs among the citizenry but also to the full satisfaction of their postmaterialist needs. Once poverty and racial discrimination were ended, and once war was all but abolished by international agreements and disarmament, Americans would enjoy an unprecedented freedom. Then the central question of American politics would become what to do with this new freedom. What was it *for*? All those promises of positive freedom that liberalism had been issuing for decades had now to be redeemed. Hence the need for a politics dealing with the insides of man, with his religious, aesthetic, cultural, moral, and spiritual quality of life, with the extent to which his way of life was fully human. What that entailed couldn't be defined by some static standard, of course, but had to be fully up-to-date. When Richard Goodwin sat down to write the Great Society speech, he knew he wanted to go beyond the liberal tradition, beyond the New Deal, which hadn't fully understood what he terms "the revelation of our modern world—that private income, a decent standard of living, was only a foundation; that private affluence, no matter how widely distributed," could not redeem the nation's spiritual emptiness. That that came as a "revelation" is a sad comment on his own spiritual condition, or on his liberal reading list. He drew on that list as he pondered his assignment, recalling in his mind Betty Friedan's *The Feminine Mystique*, Ralph Nader's writings and speeches (*Unsafe at Any Speed* came out later), and the Port Huron Statement of Students for a Democratic Society (SDS). As a progressive, he assumed that to know a time and its needs, you had to examine the present "discontents, the movements for change" in that time.[18] He figured out what Johnson should say by looking at what Johnson's critics (or figures who would soon be his critics) were saying. Johnson and the kids really were on the same page, but it was the kids' book.

One expects to find epic collisions of will and understanding between LBJ and his youthful opposition, but what's curious is how similar their visions were. With the exception of the Vietnam War, on which they did have an epic collision, the youthful protesters were often in tune with the president. They both sought "a place where the meaning of man's life matches the marvel of man's labor," though the kids were not that keen on labor maybe, and both agreed that getting to that place was only a matter of finding sufficient courage; the resources were well in hand. In his Inaugural Address, Johnson defined the Great Society as "the excitement of becoming—always becoming, trying, probing, falling, resting, and trying again—but always trying and always gaining." "Always becoming" might have been a joint motto, and his list of participles was sufficiently redolent of sex to have delighted the irreverent youth on that count, too. Yet by and large—with the passage of the Civil Rights Act of 1964 an important exception—the kids neither liked nor admired him and his achievements. He was not cool and could never become cool. With only a slight exaggeration, his White House special counsel Harry McPherson captured

LBJ's squareness: "Johnson was a manipulator of men when there was a rejection of power politics; he was a believer in institutions at a time when spontaneity was being celebrated; he was a paternalist when paternal authority was rejected; and he came to political maturity during the 1930s, when democracy was threatened by fascism and communism, making him an unbending anticommunist." Through his earnest rhetorical performances Johnson could arouse interest in an issue, but he couldn't *lead* public opinion, at least not youthful public opinion, as Jack Kennedy, FDR, and Wilson had done. He looked like an "authority figure," which had been an advantage when he was Senate leader but not as president in the 1960s. The famous Johnson "treatment" showed him at his most persuasive: up close and personal with another senator, one massive arm draped over the poor fellow's shoulder with Johnson's face just a few inches away from his.[19] Master of this style of one-on-one politics, LBJ was the opposite of Wilson's ideal leader of men, who specialized in moving mass audiences, not individuals.

Nonetheless, Johnson's victory in the 1964 election was staggering. He captured the greatest percentage of the popular vote in modern U.S. history, just over 61 percent, exceeding even FDR's peak of 60.8 percent in 1936. And his party dominated the national legislature as it hadn't since the height of the New Deal: 68 senators, and 295 members of the House (versus the Republicans' 140). It was the high-water mark of liberalism. But it was downhill from there, and quickly.

Kennedy's assassination had already created a quasi-crisis atmosphere that encouraged a kind of de facto suspension of the separation of powers, just as in Roosevelt's Hundred Days. Key legislation passed without much deliberation. (The Civil Rights Act, which passed on July 2, well before the election, was again exceptional, having been debated in the Senate's longest filibuster.) The Economic Opportunity Act, mainstay of the War on Poverty, was hardly written before it was hardly deliberated, having been thrown together in a frenzy by Johnson's advisors, trying to put flesh on the bare bones he had vouchsafed in his first State of the Union address.[20] After the election, the new Democratic superma-jorities rapidly enacted Medicare, Medicaid, the Elementary and Secondary Education Act, and a revolutionary immigration bill, among other things. Of the 87 bills submitted by the Administra-tion, 84 were enacted. The War on Poverty was under way, and the Great Society as well. On the cultural front alone, the Johnson ad-ministration added the Corporation for Public Broadcasting (estab-lishing National Public Radio and the Public Broadcasting System), the National Endowment for the Arts, and the National Endow-ment for the Humanities. With the economy roaring—gross do-mestic product grew by 7 percent in 1964, and even faster the next year—and the federal budget balanced—indeed the main budget issue was how to enlarge government spending fast enough to keep pace with growing tax revenues—it really did appear that the Great Society would be able to "do it all."

"The first sign of trouble," as Steven Hayward puts it in his fine history of "the fall of the old liberal order," came from the front lines of the War on Poverty. The Community Action Program (CAP), charged to ensure "maximum feasible participation" of the poor in their own uplifting, had gone too far, staging strikes and sit-ins, organizing tenant unions in public housing, and enlisting voters in campaigns against local mayors and political machines. "Class struggle," as the U.S. Conference of Mayors called it, was not what Johnson had had in mind, but the young antipoverty work-ers and community organizers (Saul Alinsky was a grant recipient) had their own subversive ideas. No sooner had the administration reined in CAP than even worse trouble broke out in the black com-munities of the inner cities. After the Voting Rights Act passed in 1965, the civil rights movement seemed poised to celebrate victory and go home, because in fact it had won its struggle for equality of rights for black Americans. In August, Martin Luther King Jr.

confirmed the obvious: "There is no more civil rights movement," he said. "President Johnson signed it out of existence when he signed the voting rights bill."[21]

Yet in a variety of ways Johnson, the leaders of the civil rights movement, and young blacks in the cities were all about to up the ante. At Howard University, an historically all-black school, LBJ declared on Commencement Day that "freedom is not enough. . . . You do not take a person who for years has been hobbled by chains and liberate him, bring him up to the starting line of a race and then say, 'you are free to compete with all the others,' and still justly believe that you have been completely fair." But the point of the race is to find out who is the fastest. You can't find that out by hobbling the other runners. You can only discover it by holding a fair race and affording everyone an equal start. For the Great Society, however, to equalize rights was not enough; it was necessary to equalize the exercise of rights, the outcomes. "This is the next and more profound stage of the battle for civil rights," said LBJ. The great victories of the past decade, a hundred years delayed, were but a "stage" in the fight, and a less important one at that. It was not enough to be judged by the content of one's character, as opposed to the color of one's skin. If everyone's character was not equally admirable or reliable, then there either was something wrong with the test, or further social equalization was called for. "We seek not just freedom but opportunity," Johnson continued. "We seek not just legal equity but human ability, not just equality as a right and theory but equality as a fact and equality as a result."[22] Several months later he issued an executive order commanding federal agencies "to take affirmative action" to ensure that blacks were hired. He didn't specify goals or timetables but these would follow soon enough under his successor. Equal opportunity was on the way to being replaced by equal results.[23]

The theory was that equal rights were too formal or formalistic, the same argument used by the Progressives against natural rights and even earlier by Karl Marx against "bourgeois rights." The only freedom that counted was *effective* or realized freedom. To "fulfill" these rights, in Johnson's terms, the power to exercise them had to be made equal. Yet as political scientist Harvey Mansfield notes, "A right is no longer a right if its exercise is prescribed, so as to make all rights equal in fact and all citizens equal in power." Not all speakers, for example, are equally persuasive, but if the government were to equalize the power of the speakers by specifying what is said or how it is said, the right of free speech would be no more. Moreover, the speakers would *know* that the game is fixed. Affirmative action in its eventual form of racial and gender preferences, similarly, sunders rights from the gumption or spirit required to claim them. The recipient of such affirmative action accepts his rights as compensation for his historical disadvantage or present-day incapacity, thus injuring both pride and equality.[24]

"Negro poverty is not white poverty," LBJ said famously at Howard. Due to slavery's legacy, blacks "just can not do it alone," as other American minorities have. He pointed in particular to the breakdown of the black family, for which, "most of all, white America must accept responsibility." Two months later came the Watts riots in Los Angeles. Any responsibility that whites bore for black behavior began to seem very attenuated to the white working and middle class watching those disorders, especially as the rioters shouted "burn, baby, burn" during their four-day assault on their neighborhood. When King and other civil rights leaders visited after the fires died down, they were shocked to discover how excited and empowered many of the young men on the streets felt. The nonviolent strategy was dying, disdained and rejected by the very people on whose behalf it had been faithfully adhered to for so many years. Black Power would soon be the new watchword. People who should have known better, white and black, were soon defending the rioters and even excusing surging rates of illegitimacy among black families on the grounds that it was racist and unjust to "impose middle class values" on black Americans.[25] The Great Society, not to mention the New Deal coalition, ended with a bang, followed by the long whimper of white liberal guilt.

## The Cultural Revolution

The culture wars began in the 1960s and still are very much with us. The Supreme Court and the lower federal courts deserve blame for opening many fronts in these wars, unilaterally abolishing prayer in schools, deregulating pornography, ordering busing for racial integration, and worst of all, sweeping away the abortion laws of all fifty states and ordering in their place a regime of highly permissive legalized abortion. In issuing those diktats, however, the courts were following the cultural revolution led to varying degrees by the young and the intellectual Left. When it came to changing American culture, the campuses were therefore near to the center of things, followed closely by Hollywood and the press. (The music business, which produced some great popular music in the decade, deserves treatment on its own.) The Great Society's efforts to produce greater social equality and spiritual fulfillment bent American mores in new directions as well, though often with unanticipated and unwanted consequences. Its programs opened up American politics at the state and local level to the long arm of Washington as never before, involving the federal government in everything from the construction of city parks to the curriculum and funding of local schools. Poverty was not conquered but the spirit of self-government was. For the first time, the country adopted a thoroughgoing centralization of administration, nominally culminating in President Johnson as head of the executive branch, but in fact handing vast and hard-to-account-for power to myriad experts running myriad programs in the new agencies.[26] The fiasco with CAP was a warning of things to come—of the culture war running through the Democratic Party and all the way down to the fingertips of the new federal leviathan.

The federal government was losing control of itself even as it extended its power more widely and deeply into society. To oversee the multiplying government programs (at least one per social problem, but the more the merrier) that went into "centralized administration," Congress had to decentralize itself, breaking committees into subcommittees and gradually changing itself from a deliberative body that also did government oversight into an administrative body that also legislated. Inside-the-beltway bargaining among bureaucrats, congressmen, and interest groups displaced legislating as the way the game was played, and every bargain generated new cynicism out of doors. In 1964, polls showed that 70 percent of Americans were confident that the government in Washington would do the right thing most of the time. The figure plunged in subsequent decades, and is below 20 percent today.[27] The rise of the administrative state is far from the only cause of this decline, but it is a prominent cause nonetheless. Big government is, in the end, a cynicism-generating machine, which self-declared enemies of cynicism like Barack Obama fail to understand. But then cynicism and idealism were born twins, and the cynicism with which he cut deals with the interest groups supporting his health care reform bill reveals a cynicism about cynicism that either goes all the way down in his soul or else rests ultimately on a willful idealism. Neither is a healthy basis for republican self-government.

To put it differently, there was something to the kids' disdain for what in the Sixties they called "the System." Paul Potter, a president of Students for a Democratic Society (SDS), apparently coined the term in an address to one of the first anti–Vietnam War protests, near the Washington Monument in April 1965. The war "has its roots deep in the institutions of American society," he argued, in "the system" itself. "We must name that system. We must name it, describe it, analyze it, understand it and change it." Neither he nor any of the other student radicals went very far down that list of necessary activities, as can be shown by the fact that the only name of "the system" that stuck was "the System."[28] Capital letters do not an analysis make. The Port Huron Statement, written mostly by the twenty-two-year-old Tom Hayden in 1962, contained the New Left's, that is, the student Left's, most complete self-explanation. In its opposition to "fearfulness of visions" and "refusal to hope," and its lament over "the decline of utopia and hope," it could have been written by the young Obama. It had the same dreamy impatience with egoistic individualism and longing for fraternity, honesty, and "participatory democracy." In short order Obama outgrew the Romantic belief in the latter, however, having learned that communities have to be organized by outside experts, and that politics cannot do without leaders. The New Left never quite learned that, or rather never faced the contradictions in its own positions. It wanted visions without leaders, and a stateless State oddly resembling the ideal university.

"Human independence," the statement announced, should be "the goal of man and society," which translated as "finding a meaning in life that is personally authentic." Such independence, it added quickly, did not mean the old, materialistic individualism: "the object is not to have one's way so much as it is to have a way that is one's own." As a first step toward individual and social authenticity, college students had to take the lead in "making values explicit" in order to establish "alternatives." Speaking for their generation (though no one had elected them to do so—authenticity brooks no formalities), the SDS condemned the technocratic and value-free pabulum of American universities. They were on to something, but though the kids rejected dehumanized social science they had very unclear ideas about how to establish "social goals and values." Although values are not "beyond discussion and tentative determination," the statement noted, the SDS's first task was "to convince people that ... the creation of human values is complex but worthwhile." It was a matter of *creating* values, then, not discovering them, and so the young radicals left it at proclaiming "our own social values," "our

central values," and similar ringing affirmations: "We regard men as
infinitely precious and possessed of unfulfilled capacities for reason,
freedom, and love." Caught in a confusion they did not create, they
longed for moral truth but had no idea where or how to find it. By
the 1960s, liberalism had made a crucial turn that facilitated their
dilemma—from virtues to values. The idea of objective human excellence, virtue, (and of objective human decency, rights) gave way
to "values"—the notion that all moral judgments are subjective or
relative, reflecting one's culture, emotions, or will. The term itself
first came into currency in the universities, from Max Weber's sociology and from Nietzsche's philosophy, both available in new English translations in the postwar period. By the Sixties most students
at the elite schools had learned that objective morality did not exist,
and had been tempted to flee that bleak desert to one self-created
moral mirage or another. Values may have been relative, but living
without values was impossible or at any rate ignoble. Together, the
desert and the mirages were known as existentialism.[29]

As a result, the liberal idea of infinite moral and political
progress, of an increasingly rational and just society, became untenable. History had no meaning, except what man's creative will
supplied it. It was to this collapse of liberal idealism that the radical
students were reacting. Progressivism's implosion left nothing to
redeem the moral compromises that American society had made
with racial segregation and power politics at home and abroad:
hence the all-consuming ugliness that many of the student protestors and their off-campus allies saw in America. Yet the demoral-
ization had its liberating side, too, in the sexual revolution and the
drug revolution. Both were defended as consciousness-raising, and
fun. In effect, liberalism's old goal of individual self-development
was defined downward, or upward, depending on your stage of
consciousness. Wilson's and FDR's liberalism may have looked
to a long-term cultural and moral transformation of mankind,
but practically their concern with self-development looked to the

unfolding of human talents here and now, to satisfying careers for ordinary Americans that would pay the bills and allow for music lessons and some family vacations. Sixties' liberalism focused on self-expression, and beyond that on self-creation, making your Self by choosing or willing your own values. Families dropped away, in favor of the sovereign Self creating its own lifestyle, for its own sake. The whole external world, and all standards of good and evil, dropped away. Theodore Dalrymple calls this "Real Me" ethics. Everything you do you do for the sake of the Real Me, the authentic Self behind all the conventions and "natural" distinctions. But who *is* the Real Me, exactly, who is free to be whatever? The mystery of the self-creating Self is akin to the mystery of the God who creates ex nihilo. This Self is free to create its own standards, free to satisfy its every desire, free to be a universal tyrant in its own world.[30]

But playing God is a lonely business, and so the mystery of the Real Me tended to be answered more mundanely. The authentic you is the sexual you, the ethnic you, the racial you—whatever group you identify with most passionately. In most cases, Sixties-style individual liberation elided into group liberation; e.g., civil rights devolved into Black Power, because the individual demanded recognition from his or her fellow Americans and from the State, and it was easier to obtain it as a member of a claimant group. By the 1970s, by and large, the radicals had made their peace with the State, effectively acknowledging the power of the earlier liberal argument that rights have always come to individuals through membership in a group.[31] The persistence of group rights within American liberalism is remarkable; now various kinds of identity rights joined the portfolio. To be sure, the new groups of the 1960s and 1970s were almost the opposite of those favored by Wilson and the Progressives. The Aryans were out, and blacks, ethnic minorities, and homosexuals, among others, were in. The argument was the same, but the winners and losers had switched!

An existentialist theme ran throughout the manifesto, which

led to some of its worse puerilities but also to the few critical insights it had on mainstream liberalism, which was the real target of its critique. Mario Cuomo once said that liberalism campaigns in poetry but governs in prose, and the young radicals were on to that bait-and-switch. What made the alternation between poetry and prose, or between leadership and administration, necessary was that liberalism offered ends and means that did not match up, that in fact contradicted one another. The lofty goal of complete self-development or individual spiritual fulfillment was inconsistent with the kind of social programs that the liberal State had on offer—programs administered from the top down, by accredited, nonpartisan (though often quite partisan) experts, wielding the legislative, executive, and even judicial powers of big government more or less as they saw fit.

What sustained the connection between ends and means, at least for a while, was the twin hope that (a) the bureaucrats would really be civil servants, dedicated to the public good and nothing else, and (b) that the programs would provide at least the conditions for individual fulfillment, which most people would then take advantage of as progressive beings. The highest expression of the first hope was Hegel's notion (formative for Wilson's thought) of the civil servants as "the universal class," the single class that did not have its own advantage uppermost in its mind, or that at least identified its advantage with the good of the public. Every other class—farmers, businessmen, artisans—was selfish and at any rate unschooled in thinking of the public's needs. So the State needed a well-educated, truly disinterested class to program and run the machinery of government and to inject genuine morality, objective right, into a society mostly pursuing subjective right or "egoistic individualism," as the SDS called it. But the young radicals saw through these pretensions. They had studied Hegel's two great critics on this score, and were persuaded by both of them. Marx had challenged the bureaucrats' disinterestedness, arguing that in a capitalist society they would quickly develop their own individual and collective interests and would pursue their own aggrandizement like everyone else. Max Weber, the great German sociologist, had argued that bureaucratic rule-following was an essentially soulless activity, its rationality confined to technical and instrumental questions, but on the ultimate questions of life's purpose and meaning the "experts" were wholly inexpert, condemned to be neutral and value-free.

Many members of the New Left rejected the Great Society's administrative spirit on these grounds. They thought the new programs served the System's interest, not primarily the people's; and they doubted that soulless bureaucracy could help individuals discover the authentic meaning of life. Firsthand they knew that their universities, for them the model bureaucracies, were guilty on both counts. And given the "knowledge factories" (Clark Kerr's phrase) that their universities had become, and the value-free ideal the social sciences had embraced in those days, they were right, at least up to a point. To them, Great Society–style liberalism represented precisely the kind of Authority that needed to be questioned. They realized, too, that the "Whiz Kids" who ran the Pentagon under Robert McNamara suffered from the same kind of bureaucratic self-interest and amorality that the domestic-policy bureaucrats did. That consideration encouraged them to see the war not as an anomaly or a misjudgment but as another instance of the System's standard operating procedure. The "war machine" came increasingly to darken their view of the Johnson administration, and of liberalism as a whole. The student Left thought that the Great Society was not possible *with* the Vietnam War, but not possible *without* it, either.

In a way, liberalism had always been disappointed with the American people, whose enthusiasm for fundamental reforms in society and government was at best fleeting. The small portion of young people who were the politically active "kids" felt a more sustained alienation from a country they criticized as inauthentic and unjust—a mass society and consumerist culture which to them seemed largely indifferent to racial bigotry and inequalities of power and wealth. These kids had read *The Organization Man, The Lonely Crowd, The Power Elite, Eros and Civilization, The Tragedy of American Diplomacy, The Other America,* and *One-Dimensional Man,* among many other contemporary books explaining the country's spiritual emptiness and power hunger.

The New Deal hadn't made any difference and maybe made things worse. Though regarded reverently in most quarters, liberalism's greatest political achievement so far had conspicuously failed, in the students' eyes, to lift America above bourgeois individualism and interest-group selfishness. They discerned, in effect, a problem in the relation between the first and second bills of rights. The second was supposed to have *corrected* the first as well as supplemented it, to have launched the country on a course to a less materialistic culture. Instead, the second had been absorbed by the first, or more precisely, assimilated to its spirit of private property, bourgeois rights, and conventional values. Little old ladies were not *changed* by receiving Social Security checks; the unemployed were not radicalized by their unemployment benefits; workers free to join the union did not turn syndicalist. Lenin was right: give the people rights, allow them to grow prosperous, and the Revolution was doomed. No wonder the radicals put little hope in the Great Society, even if LBJ, too, had seen some of the problems with the New Deal's legacy. The American system seemed comprehensively oppressive, most insidiously when it wasn't overtly so but tolerant in a smothering way, what Herbert Marcuse called "repressive tolerance." How could they find "meaning" in such a world? They were attracted to forms of existentialist politics not only because they lived under the existential threat of the Bomb, but because the very possibility of a meaningful universe seemed folly. They hadn't learned anything in the university, typically, to suggest otherwise. Liberalism's old-time religion, the belief in rational progress, didn't appeal because progressivism had been discredited, they thought, by the results of progress itself—the Bomb, again, not to mention the Holocaust, two world wars, Jim Crow, and now the Cold War and Vietnam.

So they had a tendency to take their feelings as their guide to ethics. "If it feels good, do it" is one well-known and hardly original side of it. But engaged by the civil rights movement, the one noble cause that fascinated them, many of the kids also concluded, "If it feels *right*, do it." Having learned that all values are relative and that reason cannot by nature tell right from wrong, they leaped to the non sequitur that feelings were a better guide than reason: if morals were grounded in emotions, then emotions can be a guide to morals. In this existential leap, the key to morality became the intensity of their passions, which was a measure of their authenticity, of the real *you* taking a stand for something or other—but at least for something rather than nothing. The love of the "demonstration," of living for the excitement or the high of the protest itself, became a characteristic part of Sixties-style self-expression.

The student radicals played at revolution, mostly, rather than seriously attempting it; the exceptions were small-timers like William Ayers. The Sixties' more enduring legacy was the strange combination, still very much with us, of a more ambitious State and a less trusted government than ever before. "Expect much and trust little," as Hugh Heclo puts it in a trenchant essay. From the Great Society we got reckless expectations of the improvements that public policy could make in American life; from the New Left and the counterculture, a bitter distrust of public men's motives. Voilá: a bigger and bigger government we trust less and less. Once the "legitimacy barrier," as James Q. Wilson called it, of the Constitution and its customs could be safely disregarded, the living constitution allowed government programs to grow and multiply almost without limit. The Great Society marked the decisive victory over that barrier. Now every social problem could be met with a government

program, subject to the whim of the voters, congressional committees, and organized interests. Heclo calls this "policy mindedness,"
the domination of politics and political discourse by issues that
demand "solutions." At the same time, the New Left taught us to
replace doubt or skepticism of political solutions, stemming from
the old-fashioned suspicion of power and worries about human
nature, with the "postmodern distrust of motive." The young radicals were always sure of their own good intentions, but of no one
else's. The Real Me is always authentic; Authority is always to be
questioned, or worse. The Movement is pure, the Establishment
corrupt—by definition.[32] Yet in the third wave of liberalism, the
liberal Establishment grew to doubt its own unselfishness, its own
disinterestedness and idealism.

The kids changed American culture and politics, though not entirely in the way they had hoped. When Richard Nixon was elected
president in 1968 and again by larger margins in 1972, radicals
and liberals alike experienced a deep crisis of confidence. History or fate seemed to have turned against them with a vengeance.
Many retreated to the academy, to Hollywood, to nonprofit foundations, and to the unelected branches of government—the courts
and the very bureaucracies the kids, at least, had scorned. Indeed,
a remarkable feature of cultural liberalism from the Sixties to the
present is that it has been pushed much more by unelected officials
than by elected ones. In keeping with this turn away from popular
government, and influenced by the New Left's swelling indictment
of America as endemically racist, sexist, homophobic, and proslavery, American liberalism itself acquired a new and persistent tone
of anti-Americanism. Progressive critics like Charles Beard used to
criticize the Constitution as un- or anti-democratic. The implication
was that the people, the overwhelming majority, were virtuous but
had been politically duped and defeated. Woodrow Wilson, Franklin Roosevelt, and Lyndon Johnson regarded Americans of the past
and present as neither better nor worse than the times allowed, but

as the world's democratic, and therefore moral, vanguard nonetheless. Out of the cultural revolution of the Sixties, however, came the sweeping dismissal of the Silent Majority, the democratic majority, American society itself as depraved—as endemically racist, sexist, imperialist, you name it. When liberals rejected the majority, the majority rejected the liberals. Liberalism lost confidence in America's past, and eventually in its own future.

None dared call it malaise. President Jimmy Carter didn't call it that, either, in his speech renowned for not invoking that word. Yet his Address to the Nation on Energy and National Goals, the malaise speech, is a fitting coda to the Great Society. The speech's goal was to revive liberalism's faith in American democracy, and American democracy's faith in liberalism. It failed. Carter had run for office in 1976 promising to be a president "who is not isolated from the people, who feels your pain, and who shares your dreams and who draws his strength and his wisdom from you." He had not stinted in his appeals to Wilsonian leadership, promising to form a government "as good as the American people," but the subtext of his campaign was to repudiate the New Left's deprecation of America as an unjust and immoral System, an indictment made more plausible by Watergate. Nonetheless, his formulation left open the crucial question, just how good *is* "the American people"? In his July 15, 1979 speech, more than halfway through his painful term of office, the answer was not so good after all.

I want to talk to you right now about a fundamental threat to American democracy. . . . The threat is . . . a crisis of confidence. It is a crisis that strikes at the very heart and soul and spirit of our national will. We can see this crisis in the growing doubt about the meaning of our own lives and in the loss of a unity of purpose for our nation. The erosion of our confidence in the future is threatening to destroy the social and the political fabric of America. . . . We've always believed in something called progress. We've always had a faith that the days of our children would be better than our own. Our people are losing that faith. . . . But just as we are losing our confidence in the future, we are also beginning to close the door on our past. In a nation that was proud of hard work, strong families, close-knit communities, and our faith in God, too many of us now tend to worship self-indulgence and consumption. Human identity is no longer defined by what one does, but by what one owns. But we've discovered that owning things and consuming things does not satisfy our longing for meaning.[33]

Was this the Great Society remonstrating with the New Left, or the New Left condemning the Great Society? It was hard to tell because they seemed parts of the same process of liberal decay. Carter had no answers to the liberals' fear that history had deserted them and consequently the country. It fell to Ronald Reagan to help restore Americans' confidence in themselves, which he did in the name not of liberalism but of conservatism, pointing not merely to America's glorious future but to her heroic past and above all to her republican principles. The liberal State endured despite President Reagan's best efforts, but liberalism seemed a spent force.

**5**

# Obama and the Crisis of Liberalism

At a press conference shortly after the 2008 election, President-elect Obama was asked if he was appointing to his administration too many Washington insiders and veterans of the Clinton administration, including Hillary Clinton as secretary of state. Not at all, he replied. The important thing was that the change the American people had voted for would be infused into his administration from the top down. *Le changement c'est moi!*—I am the Change, he said in effect. "Understand where the vision for change comes from, first and foremost. It comes from me," he explained. "That's my job, is to provide the vision in terms of where we are going, and to make sure, then, that my team is implementing." By change he meant not only the promised departures from George W. Bush's public policies, the fresh "ten-point programs" any new administration could be counted on to produce, but the *changement radical* he had promised: "fundamentally transforming the United States of America."

He didn't lack confidence, of course. "I think I'm a better speechwriter than my speech writers," he reportedly told an aide earlier that year. "I know more about policies on any particular issue than my policy directors. And I'll tell you right now that I'm . . . a better political director than my political director." Four years before, as he was beginning his epic rise, he assured a *Chicago Tribune* reporter that "I'm LeBron, baby. I can play at this level. I got game." LeBron James has plenty of game but, until 2012, no NBA championship rings, so Obama was actually underestimating himself. He's proved he has *more* game, politically speaking, than LeBron. Three years into his presidency, he agreed to be interviewed by correspondent Steve Kroft on CBS's *60 Minutes*. The television show excerpted the interview, but the whole thing appeared on the Web in a *60 Minutes Overtime* video. Left out of the broadcast interview was the President's comparison of himself with his predecessors.

> The issue here is not going to be a list of accomplishments. As you said yourself, Steve, you know, I would put our legislative and foreign policy accomplishments in our first two years against any president—with the possible exceptions of Johnson, FDR, and Lincoln—just in terms of what we've gotten done in modern history. But, you know, but when it comes to the economy, we've got a lot more work to do.

He meant to be modest, tried to be modest—that darn economy—but it just wasn't in him. With the "possible exceptions" of LBJ, FDR, and Lincoln, his first two years stack up against any president's—certainly against George Washington's or Harry Truman's or Ronald Reagan's, say. It's hardly news that politicians tend to be egotistical, but this isn't garden-variety egotism. Many conservatives would call it narcissism or even megalomania, but that's too glib. What moves Obama is closer to a form of liberal magnanimity, the desire to do, and the ability to do, the great things that history allegedly demands. He sincerely compares himself to LBJ, FDR, and Lincoln, strives to emulate them as he understands them, and seeks to display his greatness of liberal soul in his transformative speeches and actions. The surface is the key to the depths in this case, as in so many others. It may be that he falls short of his own standards, or that his standards are self-contradictory or otherwise flawed—both criticisms are true, the latter more damning. Still, it would be both uncharitable and unwise to pass over his own understanding of himself when one is trying to come to grips with his presidency and with the unhealthy state of liberalism today.[1]

Obama has a century of modern American liberalism to draw on, and in a strange way his administration has recapitulated that history. He campaigned on Hope and Change, attempting, like Woodrow Wilson (and later JFK), through soaring speechmaking to awaken the idealism of a generation and resume the forward march of progressive politics. Like FDR, Obama exploited an ongoing economic collapse to pass far-reaching regulatory reforms, boost federal stimulus spending, and enact a major new entitlement program that, not incidentally, attempted to fulfill the right to adequate medical care Roosevelt had proclaimed in 1944. And like Lyndon Johnson's administration, only much sooner, Obama faced an electoral rebellion against his signature policies that threatened to eject him and his party from power and to discredit liberalism itself. All in one term, really just two years. The acceleration and compression of events are remarkable. His administration launched a fourth wave of liberal reform to add to the storied greatness, in liberals' eyes, of the first three. But the wave crested so abruptly that it raised questions about its very existence, much less its significance. The wave was real enough in legislative and electoral terms, and coming after almost thirty years of domestic politics (and foreign policy, though that's a more complicated story) conducted in the shadow of Ronald Reagan, it surprised liberals as well as conservatives. Whether it has changed liberalism, and if so, how that change may affect liberal hopes and conservative fears, will soon become pressing questions, regardless of what happens in 2012. President Obama's tenure thus poses the test of history in concentrated form: is liberalism on its last legs, or about to be reborn?

## Sinister Visions

When Wilson injected the leadership bug into American politics, he set off far-reaching changes in the way presidents see their job and their relation to the people. In the nineteenth century, presidents rarely addressed the public except on a ceremonial occasion like Inauguration Day or informally when traveling around the country "on tour." The latter kind of speeches and even most of the inaugural addresses avoided politics in the sense of policy agendas or partisan appeals; the president talked about the nature of republican government, the genius of the Constitution, and perhaps the general goals of his administration. What we refer to as the State of the Union address, then called simply the president's Annual Message, was delivered in person by Presidents Washington and Adams, but from Jefferson through Taft was downgraded—upgraded, actually—to a written message austerely sent over to Congress. Wilson broke that tradition, among many others, and resumed the practice of delivering the message in person to a joint session of Congress. Jefferson had given up that grand oratorical occasion because to him it seemed a kingly affectation. One could say that Wilson resurrected it for the same reason, that it focused attention on the "personal force" of the man who was the president-leader. (Similarly, presidential candidates didn't appear at their party's nominating convention and so didn't give highly partisan acceptance speeches until FDR, who broke with that custom in 1932.) The numbers tell the story: Washington gave, on average, 3 informal popular speeches a year (not counting the formal Annual Message), Jefferson 5, Lincoln 16, Grant 3, and McKinley 65—the latter's verbosity probably due to the railroads' spread, which allowed for whistle-stop tours. Bear in mind, too, that many of these informal speeches were only a few minutes long. Washington's Farewell Address, by the way, was printed in the newspapers, not delivered as a speech at all. Presidents in the nineteenth century recommended policies to Congress, of course, but these proposals were conveyed via written presidential messages rather than through public speeches or press conferences. As Jeffrey Tulis writes, these customs guaranteed that "rhetoric to Congress would be *public* (available to all) but not thereby *popular* (fashioned for all)."[2]

How extraordinary the formality and chasteness of presidential rhetoric in those days appears next to the nonstop chatter of today's presidential blabbermouths. Presidents of old spoke so infrequently and guardedly out of respect for the office and for the character of republican government. They were loath to be seen as intimidating Congress, their coequal branch, by appealing over congressmen's heads to the people; they refused to compromise the constitutional independence of the executive by directly courting popularity; and fearing majority tyranny as the bane of republics, they hewed to practices and customs that nipped demagoguery in the bud. (They also lacked radio, television, and the Internet, of course, but the change in doctrine preceded the technological upgrades.) Wilson's call for the president to lead his party and the nation required that he lead public opinion, and there was no way of doing that without speaking to the public directly and passionately. And often: thus was born "the rhetorical presidency," as political scientists now call it. Teddy Roosevelt pioneered it, but Wilson gave it a theoretical justification and made it a routine part of the modern office. Liberal and conservative presidents alike have used it—Ronald Reagan was a master at appealing to the people to put pressure on their representatives. "If you can't make them see the light, you can at least make them feel the heat," he used to say. Calvin Coolidge was perhaps the only modern president to resist its blandishments out of principled taciturnity, as opposed to George H. W. Bush's temperamental reticence and sheer inarticulateness. As a result of trying to interpret and lead public opinion, presidents today have to spend a lot more time listening, or pretending to listen, to the people in town hall forums and the like, and reading public opinion polls. They also must speak so often that they need a team of speechwriters to share the work.[3]

Wilson thought the rhetorical presidency would encourage an enlightened and active citizenry, as well as facilitate the march of liberal progress. There is copious evidence of the latter, but only scattered signs of the former. Leaders seeking to move the nation must first get its attention, and the easy way to do that is by inflating every social ill into a "crisis" calling for a "solution" to the social "problem" lately become acute. "Crises give birth and a new growth to statesmanship because they are peculiarly periods of action . . . [and] also of unusual opportunity for gaining leadership and a controlling and guiding influence," noted Wilson. To overcome the separation of powers and the other constitutional stumbling blocks to precipitate action, to keep their leadership going, politicians must try to keep up "*at all times*," he emphasized, an atmosphere of crisis. The leader should exploit that atmosphere, responsibly of course, but constantly keeping in mind that "the arguments which induce popular action must always be broad and obvious arguments; only a very gross substance of concrete conception can make any impression on the minds of the masses." After interpreting the majority's felt needs and still unconscious desires, the leader must gain their confidence "by arguments which they can assimilate," appealing to "elemental" motives, "large and obvious" morality, and policies "purged of all subtlety."[4]

Hence the need for visionary rhetoric, for speeches that deal not with circumstances and constitutional logic and principle—as, say, the Lincoln-Douglas debates, Lincoln's speech at Cooper Union, his July 4, 1861 special message to Congress, and other American masterpieces of deliberative rhetoric so dealt—but with images of a more perfect future. Appealing not so much to reason as to imagination, to dreams and visions rather than premises and conclusions, the visionary address is now a liberal staple, an American staple. William Safire, the Nixon aide turned *New York Times* columnist, christened it the "I-see" speech, or part of a speech. A few examples will suffice. Here is FDR, from his Second Inaugural.

> Have we reached the goal of our vision of that 4th day of March 1933? Have we found our happy valley? I see a great nation, upon a great continent, blessed with a great wealth of natural resources. . . . I see a U.S. which can demonstrate that, under democratic methods of government, national wealth can be translated into a spreading volume of human comforts hitherto unknown, and the lowest standard of living can be raised far above the level of mere subsistence. . . . I see one-third of a nation ill-housed, ill-clad, ill-nourished. It is not in despair that I paint you that picture. I paint it for you in hope—because the nation, seeing and understanding the injustice in it, proposes to paint it out.

And this is Bill Clinton, from his Second Inaugural.

> With a new vision of government, a new sense of responsibility, a new spirit of community, we will sustain America's journey. The promise we sought in a new land we will find again in a land of new promise. In this new land, education will be every citizen's most prized possession. Our schools will have the highest standards in the world, igniting the spark of possibility in the eyes of every girl and every boy. And the doors of higher education will be open to all. The knowledge and power of the Information Age will be within reach not just of the few,

but of every classroom, every library, every child. Parents
and children will have time not only to work, but to read
and play together. . . . Our streets will echo again with the
laughter of our children, because no one will try to shoot
them or sell them drugs anymore.

Finally, here is Senator Obama in his commencement address at
Knox College in 2005.

So let's dream. . . . What if we prepared every child in
America with the education and skills they need to compete in the new economy? If we made sure that college was
affordable for everyone who wanted to go? . . . What if no
matter where you worked or how many times you switched
jobs, you had health care and a pension that stayed with
you always . . . ?
. . . Ten or twenty years down the road, that old Maytag
plant could re-open its doors as an Ethanol refinery that
turned corn into fuel. Down the street, a biotechnology research lab could open up on the cusp of discovering a cure
for cancer. And across the way, a new auto company could
be busy churning out electric cars.[5]

Despite a depressing decline in literary quality and ambition
over time—so much for progress—the passages confirm the "vision
of the future" as a powerful part of liberal rhetoric. The vision is
always of the near future, for good Wilsonian reasons as we've
seen, but all the more tempting because of its very closeness; and
it offers the next installment in straight-ahead progress with, invariably, no complications, trade-offs, unintended consequences,
or permanent obstacles posed by human nature and politics. So
Obama, for instance, can dangle before the citizens of Galesburg,
Illinois, home of Knox College, the prospect not merely of a biotech

research lab opening up down the street, but one that is on the verge of curing cancer. Bill Clinton can imagine a world in which the drug trade and presumably drug abuse are no more—not mitigated but abolished. Under the pressures of political expediency, visions of the near future elide easily into dreams of a distant future unconstrained by reality. Whatever is desirable becomes possible, to the political imagination at least; and paradoxically, the better life in America gets, the more fantastic the leader's promises must become. Increasingly, politics becomes a clash between competing visions of the future, one vision trumping another in an upward spiral of blarney. Dueling dreams don't leave much room for rational argument: you can't exactly refute a dream or disprove a vision; you're asked to believe in it as a matter of faith.[6]

Prophecies may be exploded if they don't come true, but to avoid this trap liberals try to keep their prophecies as open-ended and unspecific as possible. A prophecy may be impeached as internally inconsistent or as impossible in the real world, but then, the whole point is to transform the world as it is into the world as it should be, which means the definition of the "real world" is up for grabs. After all, progressive liberals argue that the real world is the historical world, which is still in motion, still ascending to its peak. What we have called "reality" and "human nature" in the past is a reflection of an early and inferior stage of development. Wait till you see what history has in store for mankind! FDR, arguing that Americans were still far away from their "happy valley," quoted in his Second Inaugural a famous line from the poet and musician Arthur O'Shaughnessy: "Shall we pause now and turn our back upon the road that lies ahead?" Roosevelt asked. "Shall we call this the promised land? Or, shall we continue on our way? For 'each age is a dream that is dying, or one that is coming to birth.' "[7] The verse implies that the leader-dreamer interprets or creates the dream, and the dream creates cities, empires, civilizations, a new reality.

The politics of vision is an open invitation to demagogy, and worse. Empowering the passions and the imagination to set the goals of politics, even if merely for the next stage of its advance, was always to ask for trouble, even when the leader's vision was supposed to be derived from the majority's own desires and confined by a set of highly rationalist assumptions about history's overall direction and meaning. As those assumptions began to soften and dissolve over the course of the twentieth century and especially after the Sixties, the democratic leader was left to temper his vision pragmatically, to dream of a future that "worked." But since what "worked" depended on what his goals were, and his goals depended on his values, and his values depended on his will, the leader soon found himself invited to create his own vision of the future. Even if he limited himself to willing only what the people wanted, the reason for so limiting himself grew increasingly obscure. In the end the leader, if he was philosophically au courant, would find himself willing democracy for no reason at all except that he preferred it, arbitrarily making it part of his own self-creation.

Something like liberal magnanimity was possible on the basis of the Progressive assumptions: a liberal statesman could do great deeds in furtherance of history's great ends, or end. His soul may have been a tool of history, a sensitive instrument of the Spirit of the Age, but history nonetheless needed his great passions, longings, and abilities in order to accomplish its purposes. There was something great about Wilson and FDR, including their rarefied qualities of vision and compassion, though these were far from the prideful virtues of the great-souled man celebrated in the original sense of magnanimity. Aristotle's magnanimous man possessed all the *virtues*, including preeminently the consciousness of his own moral excellence, like Churchill or Washington. The progressive statesman had all the *visions*. He didn't need George Washington's moral splendor, only energetic imaginative faculties, deep-seated sympathy with the people or the majority, and great ambition; combined, as a practical matter or perhaps as an inference from his imaginative grasp of history's possibilities, with rhetorical cunning. He was more a Caesar than a Cicero. (Adolf Berle, the Brains Truster, in his letters to Franklin Roosevelt liked to address him as "Caesar.") As a more narrow and scientific pragmatism began to affect the liberal statesman, his opportunity for greatness shriveled *pari passu*. He learned to treat life as mere "experiment." Nowadays, a truly with-it statesman who knows himself to be abandoned by nature, history, and science, and left to the creative resources of his will alone, might be obsessed with making himself great or dominant, but his acts of self-creation would have nothing at all to do with soul, much less greatness of soul. They would be eruptions of will to power.

Hanging over American liberalism, then, is the constant and perhaps increasing possibility that something very undemocratic and illiberal will come out of its impatience with constitutional forms and addiction to visionary leadership. From the beginning, liberalism sought to sap and undermine constitutional morality, the habits of mind and heart appropriate to republican government under the Constitution, and to supplant it with a new morality appropriate to a living constitution. The two overlapped, so some important pillars of the old order could be salvaged—elections, majority rule, civil rights—but even these had to be planted in new ground. To the Founders, these were aspects of or inferences from natural right. To the Progressives, these were elements of the rational State that nurtured human freedom, particularly the last and highest freedom for the full development of the human personality. So long as this historical process was thought to have ended, as we've seen, there could be final knowledge, a science, of the State and of human fulfillment. By asserting that we were close to the end but not *at* the end, and that our approach to history's end would itself be never-ending, the Progressives had to content themselves with *faith* in progress, not absolute knowledge of it. The shakier the foundations of the Progressive edifice, the tighter they clung to them. They treated faith as if it were knowledge, relying on visions to lead the way to truth. (Rational skepticism of the Socratic sort was out of the question, because it would question the root idea of progressive history, that human knowledge can be permanent and certain.) Faith in the future opened the door to doubt; and doubt of history's order, the only authority supposedly left to modern man, opened the door to nihilism.

Fortunately for America, our liberals haven't marched over the edge—even though they are very fond of marching. They haven't pursued these tendencies of thought and character to their nihilistic conclusions. Overwhelmingly, they've remained democrats and believers in Progress, but at the same time they've embraced self-expression and self-creation (one of Obama's favorite concepts), relativist pragmatism, and the multiplicity of viewpoints in our increasingly globalized, pluralistic, and multicultural society, etc., etc. Inconsistency has its advantages. But the danger of a foolish consistency is always there, especially if disenchanted liberals get egged on by the postmodernists in today's academy—or if conservative victories scare them into thinking that the Right might actually be on the right side of history. An authoritarian streak already runs through American liberalism, connecting Wilson's extreme policies at home during World War I—the repression of dissenting speech, the socialist regimentation of the economy—with Franklin Roosevelt's open invitation to Congress in his First Inaugural to give him emergency powers to handle the Depression. It's worth recalling the latter incident briefly. FDR first established the ground of the invitation:

> . . . if we are to go forward, we must move as a trained
> and loyal army willing to sacrifice for the good of a
> common discipline, because without such discipline no
> progress is made, no leadership becomes effective. We are,
> I know, ready and willing to submit our lives and property

to such discipline, because it makes possible a leadership
which aims at a larger good. . . . With this pledge taken,
I assume unhesitatingly the leadership of this great army
of our people dedicated to a disciplined attack upon our
common problems.

He then reassured his listeners that such actions were feasible
"under the form of government which we have inherited from our
ancestors," because the Constitution is so simple, practical, and
flexible. Then he drew out the implications.

It is to be hoped that the normal balance of executive
and legislative authority may be wholly adequate to meet
the unprecedented task before us. But it may be that an
unprecedented demand and need for undelayed action may
call for temporary departure from that normal balance of
public procedure.

I am prepared under my constitutional duty to recom-
mend the measures that a stricken nation in the midst of a
stricken world may require. These measures, or such other
measures as the Congress may build out of its experience
and wisdom, I shall seek, within my constitutional authority, to bring to speedy adoption.

But in the event that the Congress shall fail to take one
of these two courses, and in the event that the national
emergency is still critical, I shall not evade the clear course
of duty that will then confront me. I shall ask the Congress
for the one remaining instrument to meet the crisis—broad
executive power to wage a war against the emergency, as
great as the power that would be given to me if we were in
fact invaded by a foreign foe.[8]

He refrained from raising the logical follow-up: and if Congress *refuses* to grant me the broad executive power required by the national emergency . . . then what? In the normal understanding, Congress could not grant a president extra executive power, anyway, because he already possesses it all.[9] He spoke as if ours were a quasi-parliamentary system, in which Parliament can alter the constitution by mere statute law, offering more power to the president at will. A few whiffs of the ancient Roman dictatorship can be detected, too.[10] If that seems an overreaction, imagine it were not Roosevelt but, say, Richard Nixon who was uttering the words. At any rate, the liberal frustration with our constitutional system could only be worsened by the postmodern penchant for willful visions.

Obama is neither an old-fashioned Progressive nor a radical postmodernist. Part of what makes him interesting is how he handles the conflicting strains of his own thought. As a decent man, he believes in justice, and identifies with the civil rights movement's insistence that Jim Crow was manifestly wrong and the cause of black equality manifestly right. As a self-described progressive, he believes in change, that is, he believes that change is almost always synonymous with improvement, that history has a direction and destination, that it's crucial to be on the right side of history, not the wrong, and that it's the leader's job to discern which is the right side and to lead his people to that promised land of social equality and social justice. Yet he's skeptical of the simple-minded progressive equation of history with the inevitable triumph of justice; he fears that the foreknowledge of success or the optimistic certitude of victory would detract from the honor of standing up against Jim Crow, for example. It would also create a free rider problem: why risk opposing segregation, if its fall is inevitable? He shares the civil rights movement's sense that you have to *make* history, not just wait for it to make you. Yet if men can make history, and history makes morality, then don't human beings create their own morality? As the product of a very liberal education, alas, Obama never discovered that this quandary could be resolved by returning from history to nature as the unchanging ground of our changing experience, as the foundation of morality and politics. Returning, say, to Lincoln's and the Founders' own understanding of themselves, reconsidering their argument for the Declaration's principles, never occurred to him as a serious possibility. The progressivist assumptions, though decadent, were still too strong. He thought the only way was *forward*, which as the academy defined it meant toward postmodernism or some form of postmodern pragmatism with its denial that rational Progress or Justice exists—a possibility that he could neither fully embrace nor entirely renounce.

In his capacity as a political leader, Obama's favorite formulation is that he seeks to "shape" history. "What this generation has proved today," he declared in Iowa in 2010 after his health care victory, "is that we still have the power to shape history. In the United States of America, it is still a necessary faith that our destiny is written by us, not for us." A "necessary faith" is not necessarily true, of course; it may simply be a useful or indispensable fiction. And shaping history leaves ambiguous just how much freedom or influence human beings actually have—whether we shape history decisively or only marginally. As the victory celebration continued, he repeated, as if for emphasis, "Our future is what we make it. Our future is what we make it." That's the deeper meaning of his slogan, "Yes, we can," which he elsewhere called "a simple creed that sums up the spirit of a people." In itself the phrase sounds like a reply to "No, you can't." But was the nay-sayer denying us permission to do something, or doubting our ability to do it? If the former, "Yes, we can" is an assertion of moral right or autonomy; if the latter, it's an assertion of power or competence. For Obama, in Progressive fashion, the two appear to go together. There's no right without the social competence to realize it. Pragmatically speaking, the right and the conditions of its realization are virtually indistinguishable.

Thus it's an all-purpose phrase of righteous empowerment. Obama says, "Yes, we can" to slaves, abolitionists, immigrants, western pioneers, suffragettes, the space program, healing this nation, and repairing the world—and that's in one speech.[11] In a strange way, "Yes, we can" takes the place in his thought that "all men are created equal" held in Lincoln's thought. Insofar as it is America's national creed, it affirms that America is what we make it at any given time: America stands for the ability to change, openness to change, the willingness to constantly remake ourselves—but apparently for no particular purpose. Jon Stewart, the comedian, caught the dilemma perfectly when, joshing the president over his equivocations on the Ground Zero mosque, he said Obama's slogan, as amended, now read: "Yes, we can. . . . But should we?"

The country's saving principle, then, is openness to change. "The genius of our founders is that they designed a system of government that can be changed," Obama said in 2007 when announcing his presidential candidacy. In short, ours is the kind of country that always says, "Yes, we can" to the principle of "Yes, we can." We affirm our right to change by always changing; we shape history by reshaping ourselves. Certainly Obama reshaped his own history in his autobiographical *Dreams from My Father*, the only self-portrait by a future president to admit breezily that it would be departing from "precise chronology," making "composites" out of people he knew, changing the names of most characters, and fashioning dialogues and events to better explore his personal journey to age thirty-three.[12] It's as if Reagan had written *Dutch*, Edmund Morris's ironic biography that inserted himself as a fictional contemporary to ponder Reagan's mysterious rise. All this leaves Obama's "vision" in some doubt. The Change part is obvious, but where's the Hope? What is it we should hope *for*? He didn't disguise his agenda in 2008 or since, but to a surprising extent he emphasized the need to hope for change itself, a very stripped-down version of progressivism: Dutch had moved the country so far right that the very possibility of liberalism, of a liberal future, had to be revived. Obama's speeches during 2007 and early 2008 had a peculiarly abstract or formal quality because he was trying to exhort his followers to believe in hope itself, in the possibility of change that could lead to a change in possibilities for America. Ezra Klein got the message. After the Iowa caucuses he described it in the *American Prospect*:

> Obama's finest speeches do not excite. They do not inform. They don't even really inspire. They *elevate*. They enmesh you in a grander moment, as if history had stopped flowing passively by, and, just for an instant, contracted around you, made you aware of its presence, and your role in it. He is not the Word made flesh, but the triumph of word *over* flesh, over color, over despair.[13]

Neither Wilson nor LBJ had faced that problem. Their times were hopeful, inordinately so. Like Franklin Roosevelt in his day, however, Obama thought Americans were in a spiritual depression caused by a maldistribution of wealth and power in society, from which he had to liberate them. Economic recession and despair, which hit hard in late 2008, reinforced the people's sense of hopelessness, and strengthened Obama's rhetorical prescriptions.

For all his openness to change, there is one to which Obama consistently answers, "No, we can't." Any change that would move the country backward, in his view, is anathema. "What I'm not willing to do is go back to the days when . . . ," is a phrase that begins many a sentence in his repertory. When dealing with conservatives, his confidence in history's purpose and beneficence is miraculously raised to almost Wilsonian levels. He may not be exactly sure where history is going, but somehow he knows it's not going *there*. A certain impatience and irritability creep into his voice. If people reject his vision, he can't be a leader—and that makes it *personal.* His tone turns petulant and he begins to issue orders to follow him, damn it!

Though there are plenty of examples, the most egregious is his so-called Jobs Speech to a joint session of Congress in September 2011. He began by demanding that "we," meaning the Congress, his coequal branch, "stop the political circus" and do something to help the economy. (The Congress applauded, like the trained seals they appear at these occasions.) He proceeded to outline his American Jobs Act, which contained, he stated sternly, "nothing controversial" and was "a plan that you should pass right away." In case they missed the point, he upbraided them fifteen more times with commands to "pass this jobs bill" and heavy-handed advice to "pass it right away." "It's an outrage," he said, if they take time to deliberate over a piece of legislation that would "help" the people. Why, there are schools "throughout this country that desperately need renovating. . . . Every child deserves a great school—and we can give it to them, if we act now." And it won't cost a dime because this bill, he explained, is "paid for." He asked Congress right then and there to "increase [the] amount" of cuts that it (or its ill-named supercommittee, created as part of the debt ceiling deal) had promised to make to ease the enormous federal budget deficit, so as to cover the full cost of the Jobs Act. "Paid for," in a manner of speaking, by illusory future cuts, or if you're more cynical, by future illusory cuts. Did it occur to him to take off his shoe and pound the lectern for even greater effect? Had his speechwriters lately discovered the imperative mood, after years of living in the indicative? At any rate, it was an ugly spectacle, worse than his dressing down of the justices of the Supreme Court in his State of the Union speech the previous year.[14]

The main target of these scoldings is of course the House Republicans, who tend to obstruct his measures. But in a larger sense Obama displays the progressive impatience with politics itself. It's not merely the separation of powers, checks and balances, and other constitutional devices that often stalemate change to which liberals object. It's human nature in its present state, still so inclined to praise God rather than man, to venerate the past, and to be guided by a healthy self-love. The liberal statesman must therefore put up with a lot of "politics," in the contemptuous sense in which Obama often spits out the word: that cynical pastime of Washington interest groups, children playing jejune partisan games with laws and parliamentary rules, and visionless right-wingers trying to turn back the clock. Eventually man will be worthy of liberalism, assuming it has its way with him and conditions him to love the State as the bee loves the hive. In the meantime, it's a constant struggle to bear with this unreconstructed individualist who would rather govern his potty little self (in Chesterton's great phrase) according to his own lights than be well governed by experts for his own (purported) good.

Obama, like most liberal thinkers, dreams of overcoming man's stubbornly political nature in two ways, either by assimilating politics to the family or to the military. Chastened a little by his party's defeat in the 2010 elections and by the shootings in Tucson that wounded Representative Gabrielle Giffords, he began his 2011 State of the Union address by invoking the first theme: "We are part of the American family," and together as one we're going to "win the future," a slogan with deeply Social Darwinist roots, by the way. After the future business didn't pan out so well in numerous scrapes with the House GOP, his frustration took a different direction a year later. In his 2012 State of the Union, after celebrating Osama bin Laden's killing and the withdrawal of combat forces from Iraq, the president focused on the "courage, selflessness, and teamwork of America's armed forces." "At a time when too many of our institutions have let us down," he observed, "they exceed all expectations. They're not consumed with personal ambition. They don't obsess over their differences. They focus on the mission at hand. They work together." Hmmm . . . there was no doubt where this was going. In fact, he made it explicit. "Imagine what we could accomplish if we followed their example," he said. Yes, if politics were rigidly hierarchical, if we had to follow orders from above without question, and if living together as a free people were as unequivocal and straightforward an affair as pumping bullets into bin Laden, then we could accomplish a lot more—or a lot less, depending on how highly you value democratic self-government as an accomplishment. And the truth is that the leadership paradigm values freedom and self-rule much less than it does getting things done, attacking social problems, and making sure that liberal programs survive the struggle for existence on Capitol Hill.

Leadership is a term from the military side of politics, and one of the reasons the Founders resisted it, as we discussed in chapter 2, was their determination to preserve republican politics as a civilian forum, as the activity of a free people ruling itself. A standing army might be necessary for that people's defense, but citizens had no business longing to exchange political debate and deliberation for military solidarity and discipline. On his better days President Obama knows that, but this wasn't one of them. He went on: "When you put on that uniform, it doesn't matter if you're black or white; Asian or Latino; conservative or liberal; rich or poor; gay or straight." Nor does it matter, by the way, whether you think the war is just or unjust, prudent or imprudent. It might seem that liberals have come a long way from the protest days of the 1960s when many of them lustily denounced the American war machine; but in fact, they're still compensating or overcompensating for their contempt of the U.S. military back then. At the same time they are returning to an older Progressive tradition, highly visible in the New Deal, of trying vainly to make politics the moral equivalent of war. In any event, no one has to put on a uniform to be an equal citizen with equal rights under our Constitution.

Obama had one more riff on the Navy SEALs' killing of bin

Laden. "All that mattered that day was the mission. No one thought about politics. No one thought about themselves," though each member of the team knew "there's someone behind you, watching your back."[15] The president dwelled on the mission's execution, not its planning or the decision to authorize it—in which generals and officials in the executive branch were very definitely thinking of politics in both grand and petty senses, and rightly so. On this assurance that someone's always got your back, the military and family metaphors meet up: both compare the State to a band of brothers. Properly understood, there's some truth in the comparison. The Declaration implies that when individuals join together to make a society, they enter on the basis of one for all, and all for one, insofar as their defense against enemies external and internal is concerned. But defense against those who would deprive citizens of life, liberty, and property is not an excuse for social uniformity or socialist solidarity. In other respects citizens are free to speak, worship, and organize for private and public purposes as they see fit, so long as these are broadly consistent with private rights and the public good. It's assumed that no party or individual possesses perfect wisdom, and that in deliberation it's good to hear from many sides and mull action thoughtfully. For Obama, politics stops when a leader's vision has received the people's assent, explicit or not; then the experts, whether soldiers or civil servants, take over and implement the popular will. "These are the facts. Nobody disputes them," he told Congress curtly in the special joint session to hear another of his health care speeches, though of course millions of people disputed those "facts." Vision isn't a matter of deliberation; it's a matter of *interpretation*, of the leader's reading of unconscious or latent popular will. By choosing him over John McCain, the voters had decided whose vision is more moving and timely. No wonder he is impatient with Republicans trying to "re-litigate" his policies, especially the most important one of all, the health care bill. He confuses legislators with litigators, as though their job were not deliberative at all. The people, through him, have already issued their orders. Members of Congress now have their mission. There should be no more thought of politics or self-interest or even facts. Theirs is not to reason why, theirs is but to do or die.

## Obamacare

"I am not the first president to take up this cause," Obama told Congress in September 2009, referring to national health care reform, "but I am determined to be the last." As Harvey Mansfield pointed out, a key to Obama's political success is the way he presents himself as somehow beyond or above ordinary politics, which he disdains as a self-interested scramble. He thinks of himself, and wants us to think of him, as a nonpartisan or postpartisan figure. To the extent he must indulge in partisanship now, it's for the sake of putting an end to it in the future. "His politics is apolitical," Mansfield argued. "It considers its measures to be progressive, and progress to be irreversible."[16] In other words, Obama's postpartisanship is part of the great liberal double standard.[17] Liberals cursed by such hubris imagine they have the keys to the kingdom of History; they alone get to bless or condemn forevermore. Once the Patient Protection and Affordable Care Act passed, it instantly joined the ranks of liberal social programs that were here to stay—permanent parts of the modern State. To attempt to repeal it would be not only foolish, unheard-of, and immoral, but downright impious—that is, against the laws of history. Evildoing, in short, by evildoers, if ever there were any.

Obamacare is unpopular now, but the president is wagering that the American people never met an entitlement they didn't like *eventually*, whether out of sentiment, self-interest, or simply habit. The more one ponders his electoral, policy, and longer political agenda, the more the health care bill stands out as the centerpiece of the whole political enterprise. Stop it—repeal it, we must now say—and you have a good chance of stopping the transformation he seeks. Fail, or worse don't even try, and you permit what can be called, without exaggeration, gradual regime change at home. For the health care question involves, in its longest reach, nothing less than the form of government and the habits and character of the American people.

Obama didn't conceal the special status of the health care issue. In the speech announcing his presidential candidacy in February 2007, he vowed "that we will have universal health care in America by the end of the next president's first term." It passed in early 2010, well ahead of schedule. He said repeatedly that our existing health care system not only costs too much, but is unworthy of American ideals. "We are not a nation that lets hardworking families go without the coverage they deserve; or turns its back on those in need," he told the American Medical Association in 2009. "We are a nation that cares for its citizens. We are a people who look out for one another. That is what makes this the United States of America." Actually, every nation cares in some degree for its citizens; looking out for one another is not a distinctively American trait but a minimal part of civic friendship in any decent society. But then his point was that America, despite its citizens' remarkable individual virtues, had far to go to be a decent society. Of the myriad problems Obama wanted to tackle, health care was the biggest and the most emblematic of America's moral failings. When he addressed a joint session of Congress in September 2009, Obama quoted from Senator Teddy Kennedy's final letter to him on precisely that point. Health care reform is the " 'great unfinished business of our society,' " which is "decisive for our future prosperity"; but also, Kennedy said, it " 'concerns more than material things.' " It is " 'above all a moral issue: at stake are not just the details of policy, but fundamental principles of social justice and the character of our country.' " Obama seconded Kennedy's point enthusiastically.

From the Right and Left, critics have questioned the president's decision to spend so much of his first fourteen months on reforming health care rather than reviving the economy and restoring jobs. It was a fundamental mistake that will haunt the rest of his term, they say. A few interpret it not as a miscalculation but as a case of tunnel vision, like a pilot so obsessed with a sticky compass that he forgets to fly the plane as it heads right into a mountain. On Obama's own terms, however, the dogged persistence on health care—despite the economy, despite the plummeting polls, despite Scott Brown's election to Kennedy's place in the Senate—was progressive statesmanship of the highest order. Reforming health care was the defining issue of our time. And more important, it was the royal road to a less cruel, less selfish, less capitalist, more liberal America, and he would not abandon it or be forced off it. As he told Congress, "we did not come here just to clean up crises," even one as big as the Great Recession. "We came to build a future." And the issue "central to that future" is health care. "I understand," he confessed, "that the politically safe move would be to kick the can further down the road—to defer reform one more year, or one more election. . . . But that's not what the moment calls for." After "a century of trying," he noted, the time was ripe for health care reform. The system was at "a breaking point," reform was "a necessity we cannot postpone any longer," and "for the first time, key stakeholders [the drug companies, the American Medical Association, America's Health Insurance Plans, and so on] are aligning not against, but in favor of reform."

Which is why conservatives' resolve to repeal the health care act induces apoplexy among liberals. Conservatives are supposed to be good losers, resigned to the Left's control over the steering wheel and accelerator but cheerful about getting to apply the brakes (not too suddenly or firmly, please) in the curves. The notion that the clock *could* be turned back, that some limit to the State's growth could be discovered and enforced, that the people would hold in their hands, inspect carefully, but at last reject the Holy Grail of welfare state programs, for which liberalism has been questing, just as Obama said, *for a hundred years*—why, the liberal mind reels.[18] Even more than Reagan's victories, or Clinton's ignominious failure to pass nationalized health care, this reversal would raise doubts in the liberal mind about the liberal project. At the least, losing Obamacare after winning it could set a dangerous precedent. More immediately, it would be a serious blow to liberalism's sense of its own inevitability—the quasi-religious faith in the future so central to all progressivism, and so crucial in disarming liberalism's opponents. So conservative resistance to Obamacare must begin by confronting the historical voodoo by which liberals will try to frighten the Right into believing that resistance is futile, that repeal is doomed. These gestures are best understood as a kind of war dance, like the *haka* performed by New Zealand's rugby team before a match, designed to intimidate the opposing players. Conservatives should laugh at this attempt to get them to cooperate in their own defeat.

Of course, repeal will be difficult. President Obama can be counted on to veto any repeal legislation that reaches his desk. It will take a Republican House, Senate, and president to accomplish repeal, therefore, which means it cannot happen before 2013. Failing that, it may still be possible to choke off funding for the new health care bureaucracies, thus slowing or blocking the law's implementation. The Supreme Court had a chance to invalidate the law, but struck down only part of it. Amid all these considerations, the crucial factor will be the GOP's (and perhaps an increasing number of conservative Democrats') development of the case against Obamacare, which must then be put to the American people proudly and in broad daylight.

All along, Obama knew that selling national health care to the American people would be a delicate operation. In the primary campaign, he positioned himself to Hillary Clinton's right on the issue (he was against the individual mandate back then, before he was for it as president). Less well remembered is that he tried to sound more conservative than John McCain ("John" was going to tax your insurance benefits; Obama promised to help people afford insurance by cutting their taxes). Once he was president, Obama turned the details over to Congress. In the sixty or so presidential speeches he gave calling for health care reform, no words came more readily to his lips than "If you like your doctor, you'll be able to keep your doctor; if you like your health care plan, you'll be able to keep your health care plan." Those words were usually greeted by vigorous applause. Sometimes he was even more emphatic. In July 2009 he said, "If you've got a health care plan that you get through your employer or some other private plan, *I want you to keep it*" (emphasis added). Nothing you like about health care will change, he assured his listeners, implying that since 17 of 20 Americans like their health coverage, hardly anything would change at all—except marginally for the better: he promised to reduce premiums for the satisfied customers. The reason for his caution was obvious. The vast majority of Americans had more to lose than to gain from his plan. Hillarycare had crashed on take-off precisely because the public had come to fear the scheme's costly, painful changes. Obama could not let that happen again, so he presented his plan in the most conservative or change-averse way possible. His bold promises to fundamentally transform the country still hung in the air, but he did his best to suggest he could change the whole without affecting the majority of the parts.[19]

Obamacare's biggest changes, he promised, would be to extend the existing system's virtues to the roughly 43 million uninsured (now estimated at 50 million, due to the economic downturn), to lower costs for everyone by cutting hundreds of billions of dollars in waste, fraud, and abuse out of the system, and to prevent insurance companies from acting like insurance companies. None of these changes, he emphasized over and over, would threaten the existing health coverage of most Americans. It was the inherent implausibility of this teetering structure of promises that finally did it in, sinking the plan in the polls but not in Congress. To his Democratic allies there, Obama emphasized, on the contrary, the once-in-a-lifetime significance of the vote: this was their only chance to "meet history's test" and bring huge, fundamental change to the country.

He assured them it would be good politics, too, in 2012 and even more so in the long run. But the public is not singing "Happy Days Are Here Again," and the Democrats' latest FDR moment has yet to earn the gratitude they expected or at least hoped for. They might reflect on the differences between, say, Social Security and Obamacare. As the name suggests, the former purported to make Americans secure, or at least feel secure, in an age of economic insecurity and depression. By contrast, Obamacare makes most Americans feel *less* secure, not more. They fear, and rightly so, that it will erode the quality and accessibility of care they now enjoy, and endanger affordability and medical innovation in the future. Worse, it forces responsible people who pay for their insurance to subsidize irresponsible people who don't—thus taxing the many for the sake of the few, instead of, as FDR in effect boasted concerning the New Deal, taxing the few for the sake of the many.

Besides, Social Security is a relatively straightforward program, organized like a social insurance plan even if most recipients receive much more than they contribute, even with compound interest. But a Ponzi scheme is at least understandable. The Rube Goldberg mechanism of Obamacare is much harder to comprehend. In the first place, there are large tax increases: the Medicare payroll tax goes up and is joined by a new 3.8 percent levy on "unearned income"—for individuals earning more than $200,000 and families more than $250,000, as well as sellers of homes who earn capital gains of more than $250,000 for an individual and $500,000 for a family. None of these taxes is indexed for inflation. Eventually, so-called Cadillac insurance plans will be hit with a 40 percent excise tax. In addition, medical device makers (for example, manufacturers of wheelchairs, CT scanners, heart stents, artificial knees, and the like), drug companies, tanning salons, and insurance companies have to pay special levies.

These enormous revenues will fund the extension of health insurance or Medicaid to the uninsured (3 out of 4 of them, actually; the rest fall through the cracks) and will subsidize rates for individuals and families whose employers don't provide health coverage. The eligible policies will be sold through a set of State Exchanges called for by the act, beginning in 2014. But then to prevent insurance companies from growing rich from these new revenues, the insurers are sharply limited in their ability to raise rates or exclude coverage. Their rates will be monitored by state and federal authorities who will enforce "medical loss ratios," in effect requiring the companies to spend 80 to 85 percent of their revenues on claims regardless of the firms' administrative expenses and profits. And instead of offering a range of policies including low-cost "catastrophic" coverage, the companies will be strongly encouraged to sell to everyone comprehensive policies that incorporate the "essential benefit plan," mandating a long list of expensive services like maternity care, drug rehabilitation, and mental health treatment. Finally, insurers will be forced to issue policies to anyone regardless of how sick they are, and will be forbidden to charge sick clients more than healthy ones.

What are the act's likely consequences? These have been well studied by the clear-sighted critics who've written about Obamacare and prescribed free-market remedies for our health care problems.[20] Here is a summary of their conclusions.

The new taxes will depress the economy and discourage medical investment and innovation. The subsidies and mandates will sharply increase demand for doctors and medical services. Measures to reduce waste, fraud, and abuse—code words for cutting Medicare and Medicaid payments to doctors and hospitals—will, together with higher taxes and regulatory pressures, drive many doctors into early retirement or into another line of work. As demand surges and supply contracts, prices for medical treatment will go up, as will insurance premiums. (The tax increases on drugs, medical devices, and insurance companies will be passed on to consumers.) With government trying desperately to suppress these price spikes, shortages of medical personnel and services will occur, which will lead to long waiting lists, rationed care, and decreasing capital investment. The restrictions on insurance companies will prevent them from earning a reasonable return on investment, which will eventually drive them out of business or into the arms of the federal government. The "public option," melodramatically sacrificed by the president and the House Democrats to get this "middle-of-the-road" plan, will come back again, only this time not as an option. In short, it's very unlikely that Obamacare, as designed, will work or will work for very long without triggering a more radical crisis in American health care. As Obama said after its passage, "this isn't radical reform. . . . But it moves us decisively in the right direction."[21]

Among its other effects, the act marks a new stage in the decline of constitutional government in America. One sign of this was then House Speaker Nancy Pelosi's remark, "we have to pass the bill so that you can find out what is in it, away from the fog of the controversy." After shepherding the equally massive financial regulation bill into law, Senator Christopher Dodd was moved to say something very similar: "No one will know until this is actually in place how it works." In late August 2010, Senator Max Baucus, Finance Committee chairman, chimed in: "I don't think you want me to waste my time to read every page of the health care bill. . . . We hire experts."[22] These statements make both an epistemological and a political point. The first is that these bills are so long, complicated, and unreadable that no one who isn't an expert can possibly decipher them. That implies, in turn, that no amount or quality of democratic deliberation can clarify them to citizens, and in most cases to legislators, in advance. Indeed, Pelosi suggests that political debate itself, "controversy," mostly dims public understanding by generating "fog." The second point is that neither she nor Dodd nor Baucus is especially troubled by this breakdown in democratic accountability. With this kind of legislation, they imply, there's no choice but to trust the experts—not merely those who patch the law together, but perhaps more important, those who implement it. For the truth is that this kind of bill, more than 2,500 pages long, will mean what the bureaucrats say it means.

Or if one wants to be generous, this kind of bill will mean what the bureaucrats, in conjunction or conspiracy with their congressional overseers, say it means. In short, these bills are not so much laws as administrative to-do lists. They are contrivances fit only for the modern liberal state, ambitious to regulate all local and state affairs from the center, which means through a bureaucracy of experts. The result, again, is not a government of laws but of men, albeit men who think themselves wise. You might think that high-ranking officials like Pelosi and Baucus and Dodd would rebel at becoming appendages of the administrative machine, that, valuing their duties as lawmakers so highly, they would insist on reading and even writing the bills themselves. In the past, congressmen did rebel against, as one of them put it, becoming part of "a city council that overlooks the running of the store every day." But once the national government assumed, in political scientist John Adams Wettergreen's phrase, "responsibility for the socio-economic well-being of every American," then *somebody* had to mind that store, and Congress cut its pattern to its cloth.[23]

Obamacare creates some 159 new bureaucracies—programs, commissions, boards, and other agencies. Many are quite small, but almost every one empowers unelected officials to wield power over the future content and provision of health care. There's the Pregnancy Assistance Fund, the Elder Justice Coordinating Council, and the Cures Acceleration Network, for example. In most cases the secretary of Health and Human Services (HHS) is charged with creating these entities—part of the breathtaking power delegated to the secretary under the act's provisions. In fact, it's not so much the length of the act as its vagueness, incompleteness, and amorphousness that mark it as a newfangled administrative statute, granting power to a few to rule according to their wisdom and with very little reference to the many's consent. That is to say, the law's meaning is deliberately indeterminate, left vague so as to give maximum discretion to the unholy trinity of bureaucrats, congressional staffers, and private-sector "stakeholders" who will flesh out the act with thousands of pages of regulations (12,000 and counting so far), and then amend those as needed later on. When favored interests and constituencies want to appeal a regulatory decision, they will always find a helpful congressman ready to intervene on their behalf with the very bureaucracy he helped create.

This new kind of statute—one hates to call it law—is not meant to be "a settled, standing rule," as John Locke defined law. On the contrary, it is meant permanently to be in flux, always developing and subject to renegotiation. It is law constantly suffused with wisdom, albeit constantly changing wisdom. It is what passes for law under a living constitution. In fact, Obamacare is an excellent test case for how the original U.S. Constitution is faring against the living constitution. One implication of the latter is that the difference between constitutional and statute law tends to break down; the capital-C Constitution, the framework and limits of government, dissolves into the small-c constitution, how we govern ourselves nowadays. According to the small-c constitution, Obamacare is automatically constitutional. The only thing definitely unconstitutional would be turning back the clock—in this case, trying to repeal the collectivization of health care.

But conservatives are challenging this one-way liberal ratchet by mounting vigorous attacks on the statute's constitutionality, by the standards of the genuine Constitution. Twenty-six Republican state attorneys general sued claiming that Obamacare's individual mandate—requiring everyone to purchase health insurance, under penalty of law—exceeds Congress's power under the Commerce Clause.[24] (The Supreme Court agreed, but then changed the subject.) In addition, the act is rife with unconstitutional delegations of legislative power to executive agencies and, most flagrantly, as noted, to the secretary of HHS. Less well-known but even more ominous is the Independent Payment Advisory Board (IPAB), created to rein in Medicare spending. Its fifteen members, appointed by the president, would make recommendations to limit Medicare's budget by reducing reimbursements to doctors. Unless both houses of Congress overruled it by passing their own equal or greater cuts in Medicare, IPAB's proposals would automatically become law. What's worse, the act conspires to make IPAB permanent by mandating that no resolution to repeal it can be introduced in the legislature until January 1, 2017, or after February 1, 2017—the Constitution would be operational for one month only, and even then the repeal must pass by August 15 of that year to be valid, and would not take effect until 2020.[25] Congress could presumably unravel these restrictions and undo IPAB anytime it wanted to, unless IPAB's approval as part of the Patient Protection and Affordable Care Act was one of those "constitutional moments" that liberals like to think amend the Constitution informally, and for all practical purposes permanently. In any case, this boldface attempt to override the Constitution tells you everything you need to know about Obamacare. In the end, it will be much better for the Republic if the people, acting through their elected representatives, repudiate Obamacare rather than trusting the Court to do it. A popular repudiation would not only have greater and more enduring political impact; it would reassert the people's right to defend their own Constitution against big government and its stakeholders.

The question is whether the public can assert itself justly and intelligently in this predicament. For President Obama's efforts to transform the country are only the latest installment in modern liberalism's long-running project to change America by changing Americans' relation to their government. For a century, liberalism has argued that the old Constitution was obsolescent, an eighteenth-century artifact needing thorough modernization, which liberals were happy to supply. Key to its reform was changing the scholarly, and eventually the public, interpretation of its purpose: not to limit government, to separate powers, and to keep the national government devoted to national affairs, but to liberate government so it could grow easily, to combine powers so that experts could direct government more efficiently to good purposes, and to expand the universe of good purposes by urging the national government to take responsibility for state and local affairs all the way down to the temperature of your shower. Behind this changed view of government was a new view of rights, as we've seen. The American Founders held that human rights come from God and nature, and carried with them a religious and natural obligation to be used for the sake of human safety and happiness. We needed to form government to secure our rights, but we needed to be vigilant lest government endanger those very rights. Freedom never strayed far from the virtues essential to its purpose and to its defense. At the core of the "dependency" problem of big government is the political logic of the new economic rights: there is nothing to fear from big government because the bigger it gets, the more rights the people get. Since our rights are dependent on government, why shouldn't we be? That attitude ultimately undermines individual character and self-respect, not to mention political independence, and in America this is already a chronic problem in large sectors of the welfare state. Medicaid, food stamps, unemployment insurance, and similar programs are almost half again as generous per individual recipient as they were in 2007, and the difficulties of weaning oneself from them have grown apace. Obamacare would make these problems universal and acute.

Despite what Obama often says, the alternative to all-provident government is not selfish indifference or "the Social Darwinist idea." As observers from Alexis de Tocqueville to Marvin Olasky have pointed out, America used to have a vibrant civil society that teemed with churches, mutual insurance funds, and other voluntary associations that helped Americans take care of themselves and of one another.[26] In some ways we still do, of course, though big government has supplanted more and more of these functions. State and local governments provided another indispensable part of the social safety net. Even the Progressives' bête noire, the old-fashioned political machines like Tammany Hall, provided important welfare services. This was never a "just downright mean" country. Though it lifted the safety net to the federal level, Social Security rejected the spirit of the dole and tried to avoid the entitlement mentality by masquerading, quite effectively, as an insurance program. Its bureaucracy, devoted to cutting checks on a predictable basis, never aspired to prescribe how recipients could spend their benefits, or to rule one-sixth of the national economy.

On the moral side, the entitlement debate usually concerns the damage done when a people gets hooked on an endless supply of free goodies. Though the president's health care plan is about turning health care into just such a narcotic, it is also about the unpleasant business of coming down from the entitlement high. From the beginning, he emphasized the need to reduce the share of GDP (currently 17.9 percent) flowing to doctors, hospitals, drug companies, and insurers. His strongest argument for attending to the problem immediately was that the rising cost of health care would bankrupt the government and the country. His chief rhetorical difficulty was precisely to convey both messages—that he was turning the "right" to health care into an entitlement for everyone; and that we had to stop spending so much on health care—without confusing his audience or contradicting himself. He never quite succeeded. The majority couldn't understand how even Barack Obama could expand access to health care and contract it at the same time.

In this respect, the debate was a harbinger of many more to come about the costs of the welfare state. It exposed the dirty little secret of entitlement rights, that what the government giveth, the government taketh away. Or to put it more mildly, the rights that FDR spoke of as needing redefinition "in terms of a changing and growing social order" must also be redefined in hard times when the social order, or at least the national economy, is not growing or growing too slowly to pay for all the promises the rulers have made to the people. Consequently, the "right" to health care means also the right to limit health care. The right to "adequate medical care" implies that no one may have more than adequate (that is, excellent) care, until the "least advantaged" among us, to use John Rawls's term, have received what the government deems adequate care. Indeed, the adjustments needed to implement the right to health care may go further than that. To the contemporary liberal, doesn't the mere coexistence of adequate and excellent care imply that the Americans stuck with the former are somehow second-class citizens?

Obamacare inclines America in the long run to some combination of the following: the sullen acceptance of government-distributed scarcity; envy of people who have more than their fair share of health care; the erosion of the rule of law as politically connected patients seek special favors, and elected and unelected officials line up to grant them; and growing alienation from a system that tries to play God but does so without wisdom, justice, or mercy. These toxic sentiments will be familiar to anyone who has lived under socialism, for they are its concomitants. When added to the caustic effects of dependency on government, they amount to a prescription for an American character increasingly unfit for

self-government. As Ronald Reagan once warned, you can't social-
ize the doctors without socializing the patients.

## Obama's America

To make possible a governing liberal majority, Obama has to re-
habilitate liberalism's reputation, to separate it as much as possible
from the radical politics of the Sixties and the burden of defending
big government. President Clinton began this renewal in the 1990s.
Learning from the shambles of his own health care initiative, he
proclaimed that "the era of big government is over" and preached
the Third Way gospel of opportunity, responsibility, and community. In some ways Obama continues and sharpens Clinton's efforts, wringing all the benefits he can out of the appearance of
postpartisanship while making few sacrifices of substance. He far
outshines Clinton, however, in telling the story of America in a way
that reinforces a resurgent liberalism. More than any Democratic
president since FDR, Obama has an impressive interpretation of
American history that culminates in him, and that reworks and
counters Reagan's view of our history as the working out of American exceptionalism (including divine favor), individualism, limited
government, free market economics, and time-tested morals.

As a writer, Obama's strength is telling stories, and his account
of America is a kind of story, mixing social, intellectual, and political history. It begins with the founding—with the Declaration of
Independence and Constitution. (Not for him is Reagan's emphasis on the Puritans and their "shining city on a hill." In profound
ways, the black church replaces the Puritans in Obama's chronicle
of American spirituality.) He tries to construct a new consensus
view of the country that acknowledges, and then contextualizes,
traditional views in a way meant to be reassuring, but that points
to very untraditional conclusions. For instance, in *The Audacity
of Hope*, in a chapter titled "Values," he quotes the Declaration's
famous sentence on self-evident truths and then comments:

Those simple words are our starting point as Americans; they describe not only the foundations of our government but the substance of our common creed. Not every American may be able to recite them; few, if asked, could trace the genesis of the Declaration of Independence to its roots in eighteenth-century liberal and republican thought. But the essential idea behind the Declaration—that we are born into this world free, all of us; that each of us arrives with a bundle of rights that can't be taken away by any person or any state without just cause; that through our own agency we can, and must, make of our lives what we will—is one that every American understands.[27]

It sounds almost Lincolnian, until one notices that the rights in this "bundle" are not said to be natural, exactly, nor true, and certainly not self-evident; they are an outgrowth of eighteenth-century political thought, too recondite for most Americans to know or remember. Abraham Lincoln, when explaining the Declaration, traced its central idea to God and nature, not to eighteenth-century ideologies. He called for "all honor to Jefferson" for introducing "into a merely revolutionary document, an abstract truth, applicable to all men and all times." When Jefferson was asked about the document's source and purpose, he looked to common sense as well as to a much older and richer philosophical tradition. The Declaration placed "before mankind," not merely before Americans or Europeans, "the common sense of the subject," he wrote, and its authority rested on "the harmonizing sentiments of the day, whether expressed in conversation, in letters, printed essays, or in the elementary books of public right, as Aristotle, Cicero, Locke, Sidney, etc." A commonsense argument harmonious with the political principles of Aristotle, Cicero, Locke, and Sidney, and proceeding from an abstract truth, applicable to all men and all times, could hardly be a simple distillation of eighteenth-century ideologies—unless, of course, Jefferson and Lincoln didn't know what they were talking about. If they spoke for their age without knowing so, if they were men of their times but didn't realize it, then like their twenty-first-century countrymen they too would have been ignorant of their eighteenth-century wellsprings, but precisely because they were living in or at least not long after the eighteenth century! Jefferson's account of "the rights of man" can be argued into historical relativity only by the most impudent academic sophistry. Although "abstract" in applying to human being qua human being, regardless of race, color, sex, religion, etc., the rights mentioned in the Declaration are based on a very obvious natural distinction. Jefferson wrote of "the palpable truth, that the mass of mankind has not been born with saddles on their backs, nor a favored few booted and spurred, ready to ride them legitimately, by the grace of God." No human being may rightly treat another human being, in other words, the way any human being may rightly treat a horse or some other brute animal—buying it, selling it, working it for his own purposes. Justice is rooted in our human nature, our equality vis-à-vis one another and our inequality vis-à-vis lower and higher beings. Man is neither beast nor God, and must treat his fellow human beings accordingly. Is the difference between a human being and a horse an eighteenth-century distinction?[28]

To be sure, nothing prevents natural right from also being a characteristic element of political thinking in a particular age. In speaking of the "liberal and republican" roots of the Declaration, Obama alludes to a scholarly debate over the interpretation of the Founding that was raging when he was at Harvard Law School. Ignited by, among other works, Gordon Wood's *The Creation of the American Republic*, published in 1969, the debate challenged the prevailing view that the Founding was primarily indebted to Locke's political philosophy of economic and political liberalism (in the older sense), offering as the alternative a republican tradition stretching back from Trenchard and Gordon's *Cato's Letters* to Algernon Sidney, Harrington, Machiavelli, and ultimately Aristotle. This republicanism supposedly emphasized the common good not self-interest, citizen virtue not commerce, and represented a communitarian, anticapitalist road not taken, or perhaps a public philosophy, junior grade, that alternated or mingled with Lockeanism. The debate eventually subsided, and Wood himself came to see that he had exaggerated the distinction between the two positions. It had a great influence on the law schools, insofar as it supplied them with fresh excuses for regulating private property and a nobler pedigree than legal realism or the sociological approach. Liberals liked it because it relieved them from being so hostile to the country's beginnings. Here was a version of the Founding they could get behind, even if, as became increasingly clear, the wish for a more socialist past was father to the thought. But neither side in this debate took the question of natural right all that seriously, regarding liberalism and republicanism as contending hemispheres within the eighteenth-century ideological brain. Wood devoted no sustained attention to the Declaration, or to a thorough reading of any document from the period, preferring to assemble multifarious snippets into an encyclopedic interpretation of the era, proving what he'd assumed all along, that the political thought of the Founding was a reflection of its age.[29]

The liberal-republican debate survives in Obama's thought in his account of the American story as a blend of two themes: individualism (symbolized in the Declaration) and "unity" (symbolized in the Constitution's commitment to "a more perfect Union"). The latter phrase, plucked from the Preamble, has long been a favorite of liberals from Wilson to Bill Clinton. For Obama, unity means being your brother's and sister's keeper; it means coming together "as one American family." "If fate causes us to stumble or fall, our larger American family will be there to lift us up," he explains. In real life, he hasn't exactly been there to lift up his aunt in Boston or his hut-dwelling half brother in Kenya, but then families in real life often disappoint. Even so, the family's failings only leave more work for the State. Membership in it confers or protects our "dignity," Obama argues, in the sense of guaranteeing "a basic standard of living" and effectively sharing "life's risks and rewards for the benefit of each and the good of all." And no one can enjoy "dignity and respect" without a society that guarantees both "social justice" and "economic justice." These ramify widely, demanding, in Obama's words, that "if you work in America you should not be poor"; that a college education should be every child's "birthright"; and that every American should have broadband access. Lately, he's feeling even more generous. The "basic American promise," he said in his 2012 State of the Union address, was, and should be again, that "if you worked hard, you could do well enough to raise a family, own a home, send your kids to college, and put a little away for retirement."[30]

That sounds more like winning life's lottery than a *promise* that anyone could justly demand be fulfilled. Notice how craftily, however, Obama shifts his examples of social duty from picking up the fallen to sending someone else's kids to college. How easily liberal magicians transform needs into desires, and desires into rights. They do it right before our eyes, and never explain the secret of the trick. Still, it's revealing that he doesn't go whole hog, turning such socioeconomic goods explicitly into rights and cataloging them for our wonderment. Chastened by the right-wing and middle-class backlash against welfare rights, he follows Bill Clinton in silently recasting, say, the right to go to college on someone else's money as an "investment" in "opportunity." As Obama presents it:

> . . . opportunity is yours if you're willing to reach for
> it and work for it. It's the idea that while there are few
> guarantees in life, you should be able to count on a job that
> pays the bills; health care for when you need it; a pension

for when you retire; an education for your children that
will allow them to fulfill their God-given potential.

Actually, there are quite a few "guarantees" in a life lived in
Obama's America. Even as he's wary of rights talk after the Sixties' implosion, he also denies any fondness for "big government."
Newfangled rights would imply a big government to provide them.
He's not in favor of that; he supports "active government." And
these aren't blank-check rights because the recipient has some reciprocal responsibilities—filling out the enrollment forms, showing
up at class, making passing grades, and the like. But the obliga-
tions are usually minimal, and besides, don't responsibilities and
rights usually keep house together? So these *are* rights of a sort,
and Obama said so explicitly a month before the 2008 election in
a CNN debate with John McCain. Asked whether health care was
a privilege, a responsibility, or a right, he replied, "Well, I think it
should be a *right* for every American."[31] But he had avoided saying
so up to that point.

Obama leaves the relationship between individualism and "a
more perfect union" up in the air, to be settled pragmatically. Every
society has a similar tension between "autonomy and solidarity,"
he writes, and "it has been one of the blessings of America that the
circumstances of our nation's birth allowed us to negotiate these
tensions better than most." The circumstances, not the principles,
of our nation were key, because the wide-open continent allowed
individuals to head west and form new communities to their liking
whenever they wanted to. But the continent filled up, big corporations gradually took over from the family farm, just as Wilson and
FDR had explained generations before, and soon our "values" were
in a more serious conflict that required a bigger government to help
reconcile. Unfortunately, that government proved enduringly unpopular with conservatives, who refused to adjust to the new times;
and so finding the proper balance between the individual and the

community continues to stoke our increasingly polarized, and po-
larizing, political debates. Though he hails the Constitution as a
mechanism of "deliberative democracy," Obama doesn't mean by
that a back-and-forth on public policy conducted by the executive
and legislative branches with input from the people. Deliberation of
that kind, endorsed by *The Federalist* and consistent with natural
rights, would seek means to the ends of constitutional government.
That's too narrow for Obama, who seeks deliberation about the
ends, or at least about what our rights will be and what the Constitution should mean in the age that is dawning. He wants to turn all
of the Constitution's mechanisms—separation of powers, federalism, checks and balances—into ways of forcing a "conversation"
about our identity. In such a conversation, "all citizens are required
to engage in a process of testing their ideas against an external reality, persuading others of their point of view, and building shifting
alliances of consent."[32] Required? *An* external reality? And who
judges whether the resulting conversation meets the requirements
of democracy or not? Obama deplores the bile in our contemporary
politics, and it must puzzle him that he causes so much of it. But
he's asking for it. As Bill Buckley used to say, liberals always talk
about their tolerance and eagerness to engage with other views, but
they're always surprised to find that there *are* other views.

Obama expects twenty-first-century people to have, roughly
speaking, twenty-first-century views, as he does. What then of Jefferson and his eighteenth-century compeers? Obama soon makes
clear that despite their fine words, Jefferson and the other Founders
were less than faithful to the liberal and republican inferences of
the principles they proclaimed. Like a good law school professor,
in *The Audacity of Hope* Obama lines up evidence and argument
on both sides before concluding that, in fact, the Founders probably
did not understand their principles as natural and universal, despite
their language, but rather as confined to the white race. The Declaration of Independence "may have been," he says, a transformative

moment in world history, a great breakthrough for freedom, but "that spirit of liberty didn't extend, in the minds of the Founders, to the slaves who worked their fields, made their beds, and nursed their children." As a result, the Constitution "provided no protection to those outside the constitutional circle," to those who were not "deemed members of America's political community": "the Native American whose treaties proved worthless before the court of the conqueror, or the black man Dred Scott, who would walk into the Supreme Court a free man and leave a slave." Obama doesn't argue, as Lincoln did, that the Supreme Court majority was in error, that Dred Scott was wrongly and unjustly returned to slavery, and that Chief Justice Roger Taney's dictum—that in the Founders' view the black man had no rights that the white man was bound to respect—was a profound solecism. On the contrary, Obama accepts *Dred Scott* as rightly decided according to the standards of the time. He agrees, in effect, with Taney's reading of the Declaration and the Constitution, and with Stephen Douglas's as well. Despite his admiration for Lincoln, Obama sides with Lincoln's *opponents* in their interpretation of Jefferson and the Declaration as proslavery.[33]

Consider Obama's renowned speech on race, the one devoted to starting a national conversation on the subject and to putting the Reverend Jeremiah Wright's notorious comments in their proper context. Wright had become an issue in the campaign when videos and recordings of his sermons surfaced, showing him vigorously swearing "God damn America!" for its sins against blacks and other minorities, and arguing that the atrocities of 9/11 were a case of America's "chickens coming home to roost," payback for our imperialist and racist foreign policy in the Middle East. The dog that didn't bark on March 18, 2008, was that the crucial words "all men are created equal" do not appear in Obama's carefully composed speech. And so that "already classic address," as James Kloppenberg calls it, on a topic that Obama declared he'd been thinking about for twenty years, constitutes a very different kind of argument, with a very different view of America, than one finds in, say, Martin Luther King's great speech in 1963 at the Lincoln Memorial. Obama invokes neither Jefferson nor Lincoln. He refers to the Constitution briefly, noting its "ideal of equal citizenship" and that it "promised its people liberty, and justice, and a union that could be and should be perfected over time." But he doesn't mention the conclusion that he had announced in his book, namely, that the Declaration's and the Constitution's "people" did not include blacks, and especially not black slaves.

In short, Obama regards the original intention of both the Declaration and the Constitution to be racist and even pro-slavery. But he refrains from making the point explicit because it would *confirm* the Reverend Wright's fundamental charge, that the United States is a racist country. And the point of the speech in Philadelphia, at the National Constitution Center, close by Independence Hall, the scene of the great events of 1776 and 1787, was not merely to repeat his condemnation of Wright's remarks "in unequivocal terms" but to put the whole controversy behind him, without dwelling on his fundamental agreement with Wright's interpretation of American principles.

In truth, Obama's evaluation of Wright's statements was very equivocal. He calls the reverend's charges "not only wrong but divisive," that is, untimely, because the American people are "hungry" for a "message of unity" right now, as delivered by the junior senator from Illinois. Wright expressed "a profoundly distorted view of this country," Obama says, "a view that sees white racism as endemic, and that elevates what is wrong with America above all that we know is right with America. . . ." What that means becomes clearer a little later, when Obama declares that "the profound mistake of Reverend Wright's sermons is . . . that he spoke as if our society was static; as if no progress has been made." Yet Obama's own candidacy confirms "that America can change. That is the true genius of this nation. What we have already achieved gives us hope—the audacity to hope—for what we can and must achieve tomorrow." In blunt terms, Wright wasn't wrong that America was a racist nation, with racist principles; he was wrong, however, to speak as though the country is as racist as it used to be. "America can change," not in the sense of living up to its first principles but in the opposite sense of moving *away* from them. Except, that is, from the deepest principle of all, which expresses "the true genius of this nation"—our belief in change, or in the deliberative process that produces such change. Only the "narrative" of America, the movement away from her first principles, deserves liberals' allegiance.[34]

Wright's eruptions were dangerous to Obama not merely because they raised questions about his judgment in having Wright as his pastor, and because they raised doubts about the candidate's ability to be a unifying, postracial figure. They were dangerous above all because they represented a particularly virulent strain of the spirit of Sixties radicalism, and shook Obama's claim to have left all that behind him and behind his new movement for change. As he said in his second, more decisive repudiation of his old friend on April 29, 2008, "the reason our campaign has been so successful is because we had moved beyond these old arguments." Because he did not actually disagree with his pastor's fundamental charge but could not say so openly, Obama's reasons for denouncing Wright became oddly personal. "I don't think that he showed much concern for me," Obama told reporters. Indeed, Wright's performance at the National Press Club was "a show of disrespect to me. It's . . . also, I think, an insult to what we've been trying to do in this campaign."[35]

Obama does disagree emphatically with the Reverend Wright on the question of change. He thinks Wright is trapped in the past, even as conservatives are—two very different pasts, doubtless, but equally out of touch with the country and its future. A proper understanding of America's past—centered on change and the country's openness to it—will make sense of the present and liberate us to make a brighter, more unified future, claims Obama. His understanding of the past thus pays lip service to such things as self-evident truths, original intent, and first principles, but quickly changes the subject to values, visions, dreams, ideals, myths, and narratives. This is a postmodern "move." We can't know or share truth, postmodernists assert, because there is no truth "out there," but we can share stories, and thus construct a community of shared meaning. It's these ideas that mark his furthest departure from old-fashioned liberalism. "Usually designated by a bundle of multisyllabic terms that signal their complexity, these ideas—antifoundationalism, particularism, perspectivalism, and historicism—also decisively shaped Obama's sensibility," writes his shrewdest liberal interpreter, James Kloppenberg, who teaches American intellectual history at Harvard. He provides helpful definitions.

> By antifoundationalism and particularism I mean the denial of universal principles. According to this way of thinking, human cultures are human constructions; different people exhibit different forms of behavior because they cherish different values. By perspectivalism I mean the belief that everything we see is conditioned by where we stand. There is no privileged, objective vantage point free from the perspective of particular cultural values. By historicism I mean the conviction that all human values and practices are products of historical processes and must be interpreted within historical frameworks. All principles and social patterns change; none stands outside the flow of history. These ideas come in different flavors, more and less radical and more and less nihilist.

Kloppenberg should be praised for his candor. "Obama's sensibility, his ways of thinking about culture and politics, rests on the hidden strata of these ideas," he admits.[36]

More and less radical, more and less nihilist—Obama comes in on the "less" side, but then a little bit of nihilism goes a long way. "Implicit . . . in the very idea of ordered liberty," he writes in *The Audacity of Hope*, is "a rejection of absolute truth, the infallibility of any idea or ideology or theology or 'ism,' any tyrannical consistency that might lock future generations into a single, unalterable course, or drive both majorities and minorities into the cruelties of the Inquisition, the pogrom, the gulag, or the jihad." There is no absolute truth—and that's the absolute truth, he argues. Such feeble, self-contradictory reasoning is at the heart of Obama's very private and yet very public struggle with himself to determine whether there is anything anywhere that can truly be known, or even that is rational to have faith in. Anyone who believes, really believes in absolute truth, he asserts, is a fanatic or in imminent danger of becoming a fanatic; absolute truth is the mother of extremism everywhere. Although it's certainly a good thing that America avoided religious and political tyranny, no previous president has ever credited this achievement to the Founders' rejection of absolute truth, previously known as "truth." Is the idea that human freedom is right, slavery wrong, thus to be rejected lest we embrace an "absolute truth"? What becomes of the "universal truths" Obama himself celebrates on occasion? Surely the problem is not with the degree of belief, but with the falseness of the causes for which the Inquisition, the pogrom, the gulag, and the jihad stood. A fervent belief in religious liberty is not equivalent to a fervent belief in religious tyranny, any more than a passionate belief in democracy is equivalent to a passionate longing for dictatorship. Has he forgotten Martin Luther King's own reflections on this question in his "Letter from Birmingham Jail"? After drawing on Augustine and Thomas Aquinas to distinguish between just and unjust laws—a distinction postmodernism makes impossible to uphold except ironically—King proceeded to offer a defense against exactly the kind of charge of extremism that underlies Obama's renunciation of absolute truth.

Was not Jesus an extremist for love: "Love your enemies, bless them that curse you, do good to them that
hate you, and pray for them which despitefully use you,
and persecute you." Was not Amos an extremist for justice: "Let justice roll down like waters and righteousness
like an ever-flowing stream." . . . And Abraham Lincoln:
"This nation cannot survive half slave and half free." And
Thomas Jefferson: "We hold these truths to be self-evident,
that all men are created equal. . . ." So the question is not
whether we will be extremists, but what kind of extremists
we will be. Will be extremists for hate or for love? Will we
be extremists for the preservation of injustice, or for the
extension of justice?[37]

Nothing like the moral clarity and moral reasoning of this passage will be found in Obama's speech on race. It was hailed for
its comprehensive empathy with black Americans who've long suffered racial scorn and discrimination, as well as with working-class
whites who resent affirmative action and immigration. Predictably,
it called for a national conversation on the issue, with a view to
mutual understanding and to all factions discovering their need for
a sympathetic State to recognize their grievances. His predilection
for such conversations is the reverse side of his rejection of absolute truth. In *The Audacity of Hope*, within two pages of his criticism of the Founders for allegedly excluding black Americans from
constitutional protection as equal human beings and citizens, he
warns against all such sweeping truth claims, and indeed praises
the Founders for being "suspicious of abstraction." On every major
question in America's early history, he writes, "theory yielded to
fact and necessity. . . . It may be the vision of the Founders that
inspires us, but it was their realism, their practicality and flexibility
and curiosity, that ensured the Union's survival." Obama cannot

decide whether to blame the Founders as racists, or to celebrate them as relativists; to assail them for not applying their truths absolutely to blacks and Indians along with whites, or to praise them for compromising their too absolute principles for the sake of something concrete.[38]

His attempt to resolve this contradiction carries him into still deeper and murkier waters.

> The best I can do in the face of our history is remind myself that it has not always been the pragmatist, the voice of reason, or the force of compromise, that has created the conditions for liberty. The hard, cold facts remind me that it was unbending idealists like William Lloyd Garrison who first sounded the clarion call for justice; that it was slaves and former slaves, men like . . . Frederick Douglass and women like Harriet Tubman, who recognized power would concede nothing without a fight. It was the wild-eyed prophecies of John Brown, his willingness to spill blood and not just words on behalf of his visions, that helped force the issue of a nation half slave and half free. I'm reminded that deliberation and the constitutional order may sometimes be the luxury of the powerful, and that it has been the cranks, the zealots, the prophets, the agitators, and the unreasonable—in other words, the absolutists— that have fought for a new order.

Obama turns for inspiration to the abolitionists, drawing no distinction between a superb publicist and reasoner like Frederick Douglass and a butcher like John Brown, who was happy "to spill blood and not just words on behalf of his visions." Both were "absolutists," which by Obama's definition means they were "unreasonable," but willing to fight for "a new order." He goes on to confess he has a soft spot for "those possessed of similar certainty today," for example, the "antiabortion activist" or the "animal rights activist" who's willing to break the law. He seems to suffer from certainty envy. He respects passionate, even fanatic commitment as such. Though he may "disagree with their views," he admits that "I am robbed even of the certainty of uncertainty—for sometimes absolute truths may well be absolute." Not true, necessarily, but *absolute*. It's hard to know what he means exactly. That the "truths" are fit for the times, are destined to win out and forge a "new order"? That they are willed absolutely, not pragmatically or contingently? Even his rejection of absolute truth is now uncertain.

So, finally, in his perplexity, he turns again to Lincoln. Like "no man before or since," Lincoln "understood both the deliberative function of our democracy and the limits of such deliberation." His presidency combined firm convictions with practicality or expediency. Obama seems never to have heard of prudence, the way a statesman (and a reasonable and decent person) moves from universal principles to particular conclusions in particular circumstances. The sixteenth president, he ventures, was humble and self-aware, "maintaining within himself the balance between two contradictory ideas," that we are all imperfect and thus must reach for "common understandings," and that at times "we must act nonetheless, as if we are certain, protected from error only by providence." For a man like Lincoln there is no such thing, he says in effect, as acting with moral certainty, only acting "as if we are certain," God help us. Unlike John Brown, Lincoln was an absolutist who realized the limitations of absolutism, yet still brought forth a new order. "Lincoln, and those buried at Gettysburg," Obama concludes, "remind us that we should pursue our own absolute truths only if we acknowledge that there may be a terrible price to pay."[39] *Our own absolute truths?* Those words ought to send a shudder down Americans' constitutional spine, assuming we still have one.

## The Liberal Crisis

Liberals like crises, and one shouldn't spoil them by handing them another on a silver salver. The kind of crisis that is approaching, however, is probably not their favorite kind, an emergency that presents an opportunity to enlarge government, but one that will find liberalism at a crossroads, a turning point. Liberalism can't go on as it is, not for very long. It faces difficulties both philosophical and fiscal that will compel it either to go out of business or to become something quite different from what it has been.

For most of the past century, liberalism was happy to use relativism as an argument against conservatism. Those self-evident truths that the old American constitutional order rested on were neither logically self-evident nor true, Woodrow Wilson and his followers argued, but merely rationalizations for an immature, subjective form of right that enshrined selfishness as national morality. What was truly evident was the relativity of all past views of morality, each a reflection of its society's stage of development. But there was a final stage of development, when true morality would be actualized and its inevitability made abundantly clear, that is, self-evident. Disillusionment came, as we've seen, when the purported end or near end of history coincided not with idealism justified and realized, but with what many liberals in the 1960s, especially the young, despaired of as the infinite immorality of poverty, racial injustice, Vietnam, the System, and the threat of nuclear annihilation. Relativism rounded on liberalism. Having promised so much, liberalism was peculiarly vulnerable to the charge that the complete spiritual fulfillment it once promised was neither complete nor fulfilling. As Obama's grappling shows, intelligent and morally sensitive liberals may try to suppress or internalize the problem of relativism but it cannot be forgotten or ignored. Despite his investment in deliberative democracy, communitarianism, and pragmatic decision making, he's willing to throw it all aside at the moment of decision because it doesn't satisfy his love of justice or rather his love of a certain kind of courage or resolute action. "The blood of slaves reminds us that our pragmatism can sometimes be moral cowardice," he writes.[40] In a moment like that, a great man must follow his own absolute truth, and the rest of us are left hoping it is Lincoln and not John Brown, much less Jefferson Davis, whose will is triumphant. The great man doesn't anticipate or follow or approximate history's course; he creates it, *wills* it according to his own absolute will, not absolute knowledge.

When combined with liberalism's lust for strong leaders, this openness to Nietzschean creativity looms dangerously over the liberal future. If we are lucky, if liberalism is lucky, no one will ever apply for the position of liberal "superhero," in Michael Tomasky's term, and the role will remain vacant. But as Lincoln asked in the Lyceum speech, "Is it unreasonable then to expect, that some man possessed of the loftiest genius, coupled with ambition sufficient to push it to its utmost stretch, will at some time, spring up among us?"

> And when such a one does, it will require the people to be united with each other, attached to the government and laws, and generally intelligent, to successfully frustrate his designs.
>
> Distinction will be his paramount object; and although he would as willingly, perhaps more so, acquire it by doing good as harm; yet, that opportunity being past, and nothing left to be done in the way of building up, he would set boldly to the task of pulling down.

More worrisome even than the danger of a superman able to promise that everything desirable will soon be possible is a people unattached to its constitution and laws; and for that liberalism has much to answer.

In one crucial respect, our situation would seem more perilous than the future danger Lincoln sketched, insofar as the very definitions of political "good" and "harm" are now uncertain. Richard Rorty, the late postmodern philosopher, specialized in trying to think through the liberal dilemma, which could not be resolved but could be expressed in sharp terms. He called himself a "liberal ironist." Liberals, he said, following the political theorist Judith Shklar, are "people who think cruelty is the worst thing we do"; and an "ironist" faces up to "the contingency of his or her most central beliefs and desires," including the "belief that cruelty is horrible." Unflinching liberals understand that liberalism consists in the revulsion to human cruelty or humiliation, combined with the knowledge that hatred of cruelty is no more moral or rational than love of cruelty. (Conservatives are lovers of cruelty, he implies.) In short, thoughtful liberals recognize that liberalism is a value judgment, with no ground in truth or science or Being or anything else supposedly "out there." Its central value judgment is not even a view of justice or of nobility, exactly, or an affirmation of something good; it stands merely for the negation of cruelty. Real liberalism *is* relativism, colored by a Rousseauian pity for the suffering animal who is sensitive to humiliation. To Rorty's disappointment, most actual liberals are not relativists, of course, but he thought they may eventually fall in line and at any rate could not refute postmodernism. They cannot even change the subject, though as a practical matter he recommends this, urging them to resume fighting for continual political reforms as in the glory days, rather than listening to the siren calls of the academic and cultural Left with its endless criticisms of America and of futile efforts to reform her.[41]

Avant-garde liberalism used to be about progress; now it's about nothingness. You call that progress? Perhaps, paradoxically, that's why Obama prefers to be called a progressive rather than a liberal. It's better to believe in something than in nothing, even if the something, Progress, is not as believable as it used to be. His residual progressivism helps insure him against his instinctual postmodernism. Still, liberalism is in a bad way when it has lost confidence in its own truth, and it's an odd sort of "progress" to go back to a name it surrendered eighty years ago.

Adding to liberal self-doubt is that its monopoly on the social sciences, long since broken, has been supplanted by a multiple-front argument with conservative scholars in economics, political science, and other fields. In the beginning Progressivism commanded all the social sciences because it had invented, or imported, them all. Wilson, Franklin Roosevelt, and Lyndon Johnson could be confident in the inevitability of progress, despite temporary setbacks, because the social sciences backed them up. An expertise in administering progress existed, and experts in public administration, Keynesian economics, national planning, urban affairs, modernization theory, development studies, and a half-dozen other specialties beavered away at bringing the future to life. What a difference a half century makes. The vogue for national planning disappeared under the pressure of ideas and events. Friedrich Hayek demonstrated why socialist economic planning, lacking free-market pricing information, could not succeed. In a side-by-side experiment, West Germany far outpaced East Germany in economic development, and all the people escaping across the Wall traveled from east to west, leaving their workers' paradise behind. Keynesianism flunked the test of the 1970s' stagflation. The Reagan boom, with its repeated tax cuts, flew in the face of the orthodoxy at the Harvard Department of Economics, but was cheered by the Chicago School. Milton Friedman's advice to Chile proved far sounder than Jeffrey Sachs's to Russia. Monetarism, rational choice economics, supply-side, "government failure," "regulatory capture," "incentive effects"—the intellectual discoveries were predominantly on the right. Conservative and libertarian think tanks multiplied, carrying the new insights directly into the fray.

The scholarly counterattack proceeded in political science and the law, too. Rational choice and "law and economics" changed the agenda to some degree. Both politics and the law became increasingly "originalist" in bearing, enriched by a new appreciation for eighteenth-century sources and the original intent of the Founders and the Framers of the Constitution. Above all, the Progressives' attempt to replace political philosophy with social science foundered. After World War II, an unanticipated and at first unheralded revival of political philosophy began, associated above all with Leo Strauss, questioning historicism and nihilism in the name of a broadly Socratic understanding of nature and natural right. New studies of the tradition yielded some very untraditional results. Though there were left-wing as well as right-wing aspects to this revival, the latter proved more influential and liberating. The *unquestionability* of both progress and relativism died quietly in classrooms around the country. Economics is an instrumental science, studying means not ends, and so much of the successes of free-market economics could be swallowed, pragmatically, by liberalism's maw. The developments in political philosophy challenged the ends of progressivism, proving far more damaging to it. In sheer numbers the academy remained safely, overwhelmingly in the hands of the Left, whose members in fact grew more radical, with some notable exceptions, in these years. But they gradually lost the unchallenged intellectual ascendancy, though not the prestige, they once had enjoyed.

Thanks to this intellectual rebirth, the case against Progressivism and in favor of the Constitution is stronger and deeper than it has ever been. Progressivism has never been in a fair fight, an equal fight, until now, because its political opponents had largely been educated in the same ideas, had lost touch, like Antaeus, with the ground of the Constitution in natural right, and so tended to offer only Progressivism Lite as an alternative. The sheer superficiality of Progressive scholarship is now evident. They could never take the ideas of the Declaration and Constitution seriously, for many of the same reasons that Obama cannot ultimately take them seriously. Wilson never demonstrated that the Constitution was inadequate to the problems of his age—he asserted it, or rather assumed it. His references to *The Federalist* are shallow and general, never betraying a close familiarity with any paper or papers, and willfully ignorant of the separation of powers as an instrument to energize and hone, not merely limit, the national government. Like many of his contemporaries, his criticisms of the national government are based on an exaggeratedly negative reading of constitutional theory and practice, as though John C. Calhoun had been right to see it as a weak compact, devoted above all to inaction lest action harm the propertied interests. Though he thought of himself as picking up where Hamilton, Webster, and Lincoln had left off, Wilson never investigated where they left off and why. Neither he nor his main contemporaries asked how far *The Federalist*'s or Lincoln's reading of national powers and duties might take them, because they assumed it would not take them very far, that it reflected the political forces of its age and had to be superseded by new doctrines for a new age. They weren't interested in Lincoln's reasons, only in his results. Not right but historical might was the Progressives' true focus.

Today liberalism looks increasingly, well, elderly. Hard of hearing, irascible, enamored of past glories, forgetful of mistakes and promises, prone to repeat the same stories over and over—it isn't the youthful voice of tomorrow it once imagined itself to be. Only a rhetorician of Obama's youth and artfulness could breathe life into the old tropes again. Even he can't repeat the performance in 2012. With a track record to defend, he will have to speak more prose and less poetry. With a century-old track record, liberalism will find it harder than ever to paint itself as the disinterested champion of the public good. Long ago it became an Establishment, one of the estates of the realm, with its court-party of notoriously self-interested constituencies, the public employee unions, the trial lawyers, the feminists, the environmentalists, and the corporations aching to be public utilities paying private-sector salaries. Not visions of the future, but visions of plunder come to mind. This is one side of what Walter Russell Mead means when he criticizes the "blue state social model" as outmoded and heavy-handed.[42] The Patient Protection and Affordable Care Act is about as sleek and innovative as the several phone books' worth of paper it takes up in printed form. Can one imagine Steve Jobs's reaction if he had been tasked with reading, much less implementing, the PPACA? This top-down bureaucratic monstrosity could have been written by the faculty of the Rexford G. Tugwell School of Public Administration in 1933. In fact it resembles Roosevelt's NIRA (the National Industrial Recovery Act) in its attempt to control a huge swath of the economy through collusive price-fixing, restraints on production, aversion to competition, and corporatist partnerships between industry and government. It is exhibit A in the case for the intellectual obsolescence of liberalism.

Finally, we come to the fiscal embarrassments confronting contemporary liberals. Again, Obamacare is wonderfully emblematic. President Obama's solution to the problem of two health care entitlement programs quickly going bankrupt—Medicare and Medicaid—is to add a *third*? Perhaps it is a stratagem, as we discussed before. More likely it is simply the reflexive liberal solution to any social problem: spend more. From Karl Marx to John Rawls, if you'll excuse the juxtaposition, left-wing critics of capitalism have often paid it the supreme compliment of presuming it so productive an economic system that it has overcome permanently the problem of scarcity in human life. Capitalism has generated a "plenty." It has distributional problems, which produce intolerable social and economic instability; but eliminate or control those inconveniences and it could produce wealth enough not only to provide for every man's necessities but also to lift him into the realm of freedom. To some liberals, that premise implied that socioeconomic rights could

be paid for without severe damage to the economy, and without oppressive taxation at least of the majority. Obama is the first liberal
to suggest that even capitalism cannot pay for all the benefits promised by the American welfare state, particularly regarding health
care. Granted, his solution is counterintuitive in the extreme, which
makes one wonder if he is sincere. To the extent that liberalism
is the welfare state, and the welfare state is entitlement spending,
and entitlements are mostly spent effecting the right to health care,
the insolvency of the health care entitlement programs is rightly
regarded as a major part of the economic, and moral, crisis of liberalism. "Simply put," Yuval Levin writes, "we cannot afford to
preserve our welfare state in anything like its present form." According to the Congressional Budget Office, by 2025 Medicare,
Medicaid, Social Security, and the interest on the federal debt will
consume all—all—federal revenues, leaving defense and all other
expenditures to be paid for by borrowing; and the debt will be approaching twice the country's annual GDP.[43]

If something can't go on forever, Herbert Stein noted sagely, it
won't. It would be possible to increase federal revenues by raising
taxes, but the kind of money that's needed could only be raised
by taxing the middle class (defined, let us say, as all those families
making less than $250,000 a year) very heavily. Like every Demo-
cratic candidate since Walter Mondale, who made the mistake of
confessing to the American people that he was going to raise their
taxes, Obama swore not to do that. Even supporters of Obamacare,
like Clive Crook, a commentator for the *Atlantic* and the *Financial
Times*, regretted the decision.

It is right to provide guaranteed health insurance, but
wrong to claim this great prize could be had, in effect, for
nothing. Broadly based tax increases and fundamental
reform to health care delivery will be needed to balance the
books. Denying this was a mistake. What was worse—an

insult to one's intelligence, really—was to argue as Obama
has . . . that this reform was, first and foremost, a cost-
reducing initiative, and a way to drive down premiums.[44]

If the bankruptcy of the entitlement programs were handled
just the right way, with world-class cynicism and opportunism, in
an emergency demanding quick, painful action lest grandma de-
scend into an irreversible diabetic coma, then liberalism might succeed in maneuvering America into a Scandinavia-style überwelfare
state, fueled by massive and regressive taxes cheerfully accepted
by the citizenry. But odds are we stand, instead, at the twilight
of the liberal welfare state. As it sinks, a new, more conservative
system will likely rise that will feature some combination of more
means-testing of benefits, a switch from defined-benefit to defined-
contribution programs, greater devolution of authority to the states
and localities, a new budget process that will force welfare expenditures to compete with other national priorities, and the redefinition of the welfare function away from fulfilling socioeconomic
"rights" and toward charitably taking care of the truly needy as
best the community can afford, when private efforts have failed
or proved inadequate. Currently, the welfare state operates almost
independently alongside the general government. Taken together,
these reforms will work to reintegrate the welfare state into the government, curtailing its state-within-a-state status, and even more
important, integrating it back into the constitutional system that
stands on natural rights and consent.

Is it just wishful thinking to imagine the end of liberalism? Few
things in politics are permanent. Conservatism and liberalism didn't
become the central division in our politics until the middle of the
twentieth century. Before that American politics revolved around
such issues as states' rights, wars, slavery, the tariff, and suffrage.
Parties have come and gone in our history. You won't find many
Federalists, Whigs, or Populists lining up at the polls these days.

Britain's Liberal Party faded from power in the 1920s. The Canadian Liberal Party collapsed in 2011. Recently, within a decade of its maximum empire at home and abroad, a combined intellectual movement, political party, and form of government crumbled away, to be swept up and consigned to the dustbin of history. Communism, which in a very different way from American liberalism traced its roots to Hegel, Social Darwinism, and leadership by a vanguard group of intellectuals, vanished before our eyes, though not without an abortive coup or two. If communism, armed with millions of troops and thousands of megatons of nuclear weapons, could collapse of its own deadweight and implausibility, why not American liberalism? The parallel is imperfect, of course, because liberalism and its vehicle, the Democratic Party, remain profoundly popular, resilient, and changeable. Elections matter to them. What's more, the egalitarian impulse, centralized government (though not centralized administration), and the Democratic Party have deep roots in the American political tradition—and reflect permanent aspects of modern democracy itself, as Tocqueville testifies.

Some elements of liberalism are inherent to American democracy, then, but the compound, the peculiar combination that is contemporary liberalism, is not. Compounded of the philosophy of history, Social Darwinism, the living constitution, leadership, the cult of the State, the rule of administrative experts, entitlements and group rights, and moral creativity, modern liberalism is something new and distinctive, despite the presence in it, too, of certain American constants like the love of equality and democratic individualism. Under the pressure of ideas and events, that compound could come apart. Liberals' confidence in being on the right, the winning side of history could crumble, perhaps has already begun to crumble. Trust in government, which really means in the State, is at all-time lows. A majority of Americans opposes a new entitlement program—in part because they want to keep the old programs unimpaired, but also because the economic and moral sustainability of the whole welfare state grows more and more doubtful. The goodwill and even the presumptive expertise of many government experts command less and less respect. Obama's speeches no longer send the old thrill up the leg, and his leadership, whether for one or two terms, may yet help to discredit the respectability of following the Leader.

The Democratic Party is unlikely to go poof, but it's possible that modern liberalism will. A series of nasty political defeats and painful repudiations of its impossible dreams might do the trick. At the least, it will have to downsize its ambitions and get back in touch with political, moral, and fiscal reality. It will have to—all together now—*turn back the clock*. Much will depend, too, on what conservatives say and do in the coming years. Will they have the prudence and guile to elevate the fight to the level of constitutional principle, to expose the Tory credentials of their opponents? President Obama's decision to double down aggressively on the reach and cost of big government, just as the European model of social democracy is hitting the skids, provides the perfect opportunity for conservatives to exploit. His course makes the problems of liberalism worse and more urgent, as though he is eager for a crisis. Sooner or later, the crisis will come. If the people remain attached to their government and laws, and American statesmen do their part, the country may yet take the path leading up from liberalism.

# Acknowledgments

**W**hen Adam Bellow, the editor of Broadside Books at Harper-Collins Publishers, called to propose that I write this book, I readily agreed. I had been thinking about the subject for a long time. My thanks to him for making that call, and for his and associate editor Kathryn Whitenight's patient ministrations in bringing it to print.

I began writing it while on a sabbatical from Claremont McKenna College that the Dean of Faculty, Gregory Hess, was instrumental in arranging. I'm grateful to him, and to Mark Blitz, the director of CMC's Henry Salvatori Center for the Study of Individual Freedom in the Modern World, who not only funded a research assistant for me but also read an early version of the manuscript. I appreciate, too, the many Claremont students in myriad courses over the years who listened to, occasionally rebelled against, and frequently clarified my arguments.

Kathleen Arnn provided the expert and enthusiastic research help, for which I'm deeply indebted. Christopher Flannery and John Kienker read the book in its final stages and supplied good cheer and good counsel, both much appreciated. Patrick Collins did a ruthless fact-checking job, as ever. And special thanks to the president of the Claremont Institute for the Study of Statesmanship and Political Philosophy, Brian Kennedy, for his long-standing support and friendship.

My deepest gratitude is reserved for my wife, Sally Pipes, who encouraged me throughout, and whose love and inspiration made the work light, and the time fly.

# Endnotes

**Introduction**

1. Stanley Kurtz, *Radical-in-Chief: Barack Obama and the Untold Story of American Socialism* (New York: Threshold Editions, 2010), pp. 1–11, 21-60, 71–77, 86.

2. For an account of the Left that says very little about the political mainstream, see Michael Kazin, *American Dreamers: How the Left Changed a Nation* (New York: Knopf, 2011).

3. Cf. Will Morrisey, "Theodore Roosevelt on Self-Government and the Administrative State," in John Marini and Ken Masugi, eds., *The Progressive Revolution in Politics and Political Science: Transforming the American Regime* (Lanham: Rowman & Littlefield, 2005), ch. 2.

**Chapter 1: The Audacity of Barack Obama**

1. See Edward C. Banfield, *The Unheavenly City* (Boston: Little, Brown, 1970), chapter 9.

2. Todd Gitlin, *The Sixties: Years of Hope, Days of Rage* (New York: Bantam Books, 1987), pp. 319, 333–34. According to Gitlin, "Mayor Daley yelled something which the TV sound couldn't pick up but lip-readers later decoded as 'Fuck you you Jew son of a bitch you lousy motherfucker go home.'" Norman Mailer, for once, was more discreet: "Daley seemed to be telling Ribicoff to go have carnal relations with himself." See his *Miami and the Siege of Chicago: An Informal History of the Republican and Democratic Conventions of 1968* (New York: World, 1968), pp. 179–81.

3. Haynes Johnson, "1968 Democratic Convention: The Bosses Strike Back," *Smithsonian* (August 2008), www.smithsonianmag.com/history-archaeology/1968-democratic-convention.html.

4. Gitlin, *The Sixties*, pp. 285–94, 341–48.

5. Nancy Gibbs, "How Obama Rewrote the Book," *Time*, November 5, 2008, www.time.com/time/magazine/article/0,9171,1856982,00.html.

6. David Remnick, *The Bridge: The Life and Rise of Barack Obama* (New York: Knopf, 2010), pp. 556–60.

7. Michael Tomasky, "Against Despair," *Democracy: A Journal of Ideas*, no. 17 (Summer 2010), pp. 57–70, at 68.

8. Ibid., pp. 68, 70.

9. Obama won 52.9 percent of the popular vote, Jackson 56.1 percent, and FDR 57.4 percent. LBJ garnered 61.1 percent in his 1964 victory. Strictly speaking, the numbers for Jackson are not comparable, because two of the states did not hold a popular vote for president in 1828. Delaware and South Carolina left it up to the state legislature to assign the state's votes in the Electoral College. See Gerhard Peters, "Presidential Election

Mandates," in *The American Presidency Project*, ed. John T. Woolley and Gerhard Peters (Santa Barbara: University of California, 1999–2009), available at http://www.presidency.ucsb.edu/data/mandates.php.

10. For example, the American Recovery and Reinvestment Act avoided a Senate filibuster, passing 61–37; the Patient Protection and Affordable Care Act lost 34 Democratic votes but still passed in the House, 219–212; and the Dodd-Frank Wall Street Reform and Consumer Protection Act, shunned by 27 Democrats, nonetheless cleared the House with a 223–202 vote.

11. Barack Obama, *The Audacity of Hope: Thoughts on Reclaiming the American Dream* (New York: Three Rivers Press, 2006), pp. 36–37, 104, 107–8, 265, 327, 354–55.

12. Ibid., pp. 34–36.

13. Barack Obama, "The American Promise": Address Accepting the Presidential Nomination at the Democratic National Convention in Denver, August 28, 2008, http://www.presidency.ucsb.edu/ws/index.php?pid=78284&st=&st1=#axzz1lvulJr36.

14. Barack Obama, "In Their Own Words: Obama on Reagan," *New York Times*, http://www.nytimes.com/ref/us/politics/21seelye-text.html.

15. James W. Ceaser, "What a Long Strange Race It's Been," *Claremont Review of Books* 8, no. 2 (Spring 2008), pp. 8–11.

16. Mark Morford, "Is Obama an Enlightened Being? Spiritual Wise Ones Say: This sure ain't no ordinary politician. You buying it?" *San Francisco Gate*, June 6, 2008, http://articles.sfgate.com/2008-06-06/entertainment/17120245_1_obama-s-presence-new-age-black-president; Barack Obama, *The Audacity of Hope*, pp. 31–33, 53, 97–98, 153–57, 176–78, 193, 288–89, 361–62.

17. Remarks by Senator Barack Obama at the Opening of the Abraham Lincoln Presidential Library and Museum, April 20, 2005, http://chicago.about.com/od/chicagopeople/a/ObamaSpeechAL.htm; Michelle Obama, Speech at the University of California, Los Angeles, February 1, 2008, http://fedpapers.blogspot.com/2008/02/michelle-obamas-speech-ucla-two-weeks.html; Lauren Collins, "The Other Obama," *New Yorker*, March 10, 2008, http://www.newyorker.com/reporting/2008/03/10/080310fa_fact_collins?printable=true&currentPage=all; William Kristol, "It's All About Him," *New York Times*, February 25, 2008, http://www.nytimes.com/2008/02/25/opinion/25kristol.htmlref=opinion.

18. Barack Obama, Speech at the Democratic National Convention, July 27, 2004, http://www.americanrhetoric.com/speeches/convention2004/barack-obama2004dnc.htm; Speech before Planned Parenthood Action Fund, July 17, 2007, http://sites.google.com/site/lauraetch/barackobamabeforeplanned-parenthoodaction; Remarks to the California State Democratic Convention in San Diego, May 2, 2007, http://www.presidency.ucsb.edu/ws/index.php?pid=77041&st=&st1=#axzz1lvulJr36; "The American Promise": Address Accepting the Presidential Nomination at the Democratic National Convention in Denver, August 28, 2008, http://www.presidency.ucsb.edu/ws/index.php?pid=78284&st=attack&st1=#axzz1j058LbOV; Remarks Announcing

Candidacy for President in Springfield, Illinois, February 10, 2007, http://
www.presidency.ucsb.edu/ws/index.php?pid=76999&st=&st1=#axzz1lvu
lJr36.

19. Obama, *The Audacity of Hope*, pp. 32–34, 37–40.

20. Obama, Remarks Announcing Candidacy for President in Springfield, Illinois, February 10, 2007, http://www.presidency.ucsb.edu/ws/index.php?
pid=76999&st=&st1=#axzz1lvulJr36; Remarks Following the Wisconsin
Primary, February 19, 2008, http://www.presidency.ucsb.edu/ws/index.php
?pid=76558&st=&st1=#axzz1lvulJr36.

21. Barack Obama, Remarks Announcing Candidacy for President in Springfield, Illinois, February 10, 2007, http://www.presidency.ucsb.edu/ws/index.
php?pid=76999&st=&st1=#axzz1lvulJr36; Remarks Following the "Super
Tuesday" Primaries and Caucuses, February 5, 2008, http://www.presi-
dency.ucsb.edu/ws/index.php?pid=76361&st=&st1=#axzz1lvulJr36.

22. Obama, *The Audacity of Hope*, pp. 356–57; Speech at the Democratic
National Convention, July 17, 2004, http://www.americanrhetoric.com/
speeches/convention2004/barackobama2004dnc.htm; Remarks Following
the Wisconsin Primary, February 19, 2008, http://www.presidency.ucsb.
edu/ws/index.php?pid=76558&st=&st1=#axzz1lvulJr36.

23. Aristotle, *Nicomachean Ethics*, trans. Robert Bartlett and Susan Collins (Chicago: University of Chicago Press, 2011), Book 7, Chapter 14,
1154b10–15, p. 161; Thomas Aquinas, *Summa Theologica*: I-II, Q. 40, 1,
6, http://www.newadvent.org/summa/2040.htm; II-II, Q. 17, 1, 5–6, at
http://www.newadvent.org/summa/3017.htm; II-II, Q. 127, 1–2, http://
www.newadvent.org/summa/3127.htm.

24. Barack Obama, Address Before a Joint Session of the Congress on Health
Care Reform, September 9, 2009, http://www.presidency.ucsb.edu/ws/
index.php?pid=86592&st=&st1=#axzz1lvulJr36; Remarks Announcing
Candidacy for President in Springfield, Illinois, February 10, 2007, http://
www.presidency.ucsb.edu/ws/index.php?pid=76999&st=&st1=#axzz1lv
ulJr36; Remarks Following the Iowa Caucuses, January 3, 2008, http://
www.presidency.ucsb.edu/ws/index.php?pid=76232&st=&st1=#axzz1lvu
lJr36; Remarks Following the New Hampshire Primary, January 8, 2008,
http://www.presidency.ucsb.edu/ws/index.php?pid=62272&st=&st1=#axzz
1lvulJr36; Remarks in Columbia, Missouri, October 30, 2008, http://www.
presidency.ucsb.edu/ws/index.php?pid=84665&st=&st1=#axzz1lvulJr36.

25. See Johan Norberg, "The Great Debt Bubble of 2011," *Spectator*, January
1, 2011, pp. 12–13.

26. William Voegeli, *Never Enough: America's Limitless Welfare State* (New
York: Encounter Books, 2010), p. 154.

## Chapter 2: Woodrow Wilson and the Politics of Progress

1. Barack Obama, Address Before a Joint Session of the Congress on Health
Care Reform, September 9, 2009, http://www.presidency.ucsb.edu/ws/
index.php?pid=86592&st=&st1=#axzz1lvulJr36; Remarks in Columbia,
Missouri, October 30, 2008, http://www.presidency.ucsb.edu/ws/index.
php?pid=84665&st=&st1=#axzz1lvulJr36; Remarks Following the Iowa

Caucuses, January 3, 2008, http://www.presidency.ucsb.edu/ws/index.php?
pid=76232&st=&st1=#axzz1lvulJr36.

2.  Woodrow Wilson, *The Papers of Woodrow Wilson*, ed. Arthur S. Link, vol. 5
(Princeton, NJ: Princeton University Press, 1968), pp. 389–90, and vol. 2
(Princeton, NJ: Princeton University Press, 1967), p. 500; see also vol. 5,
pp. 328–46. Cf. Paul Eidelberg, *A Discourse on Statesmanship: The Design
and Transformation of the American Polity* (Urbana: University of Illinois
Press, 1974), pp. 315–16.

3.  Contrast, for example, their views on the president's role in making treaties,
a topic highly relevant to the Senate's failure in 1919 to ratify the Treaty of
Versailles—a crushing blow to Wilson's presidency. Hamilton cautioned:
"The history of human conduct does not warrant that exalted opinion of
human virtue which would make it wise in a nation to commit interests of
so delicate and momentous a kind, as those which concern its intercourse
with the rest of the world, to the sole disposal of a magistrate created and
circumstanced as would be a President of the United States." Alexander
Hamilton, James Madison, and John Jay, *The Federalist Papers*, ed. Charles
R. Kesler (New York: Signet Classics, 2003), no. 75, p. 450. By contrast,
Wilson boasted of the president's "control, which is very absolute, of the
foreign relations of the nation. The initiative in foreign affairs, which the
President possesses without any restriction whatever, is virtually the power
to control them absolutely. The President cannot conclude a treaty with a
foreign power without the consent of the Senate, but he may guide every
step of diplomacy, and to guide diplomacy is to determine what treaties must be made. . . . He need disclose no step of negotiation until it is
complete, and when in any critical matter it is completed the government
is virtually committed." Woodrow Wilson, *Constitutional Government in
the United States* (New York: Columbia University Press, 1911), pp. 77–78.
Cf. Jeffrey K. Tulis, *The Rhetorical Presidency* (Princeton, NJ: Princeton
University Press, 1987), pp. 157–61.

4.  Woodrow Wilson, *The New Freedom: A Call for the Emancipation of the
Generous Energies of a People* (New York: Doubleday, Page, 1913), p. 40.
But cf. Herbert Croly, *The Promise of American Life* (Cambridge, MA:
Belknap Press, 1965; orig. ed., 1909), pp. 167–75. The best accounts of TR
as a political thinker are Jean Yarbrough, *Theodore Roosevelt and the Pro-
gressive Critique of the Founding* (Lawrence: University Press of Kansas,
2012); and Lance Robinson, "Theodore Roosevelt and William Howard
Taft: The Constitutional Foundations of the Modern Presidency," in *The
Constitutional Presidency*, ed. Joseph M. Bessette and Jeffrey K. Tulis (Baltimore: Johns Hopkins University Press, 2009), pp. 76–95.

5.  Richard Hofstadter, *The American Political Tradition and the Men Who
Made It* (New York: Vintage, 1989) p. 306ff.; Sidney Milkis, *Theodore
Roosevelt, the Progressive Party, and the Transformation of American
Democracy* (Lawrence: University Press of Kansas, 2011).

6.  See Louis Auchincloss, *Woodrow Wilson: A Life* (New York: Penguin
Books, 2009), pp. 4–6; Eric Foner, *The Story of American Freedom*
(New York: Norton, 1998), p. 186. On Progressivism and the pro-slavery

argument, consider C. Edward Merriam, *A History of American Political Theories* (New York: Macmillan, 1920), pp. 307 and 312. Cf. Henry Blumenthal, "Woodrow Wilson and the Race Question," *Journal of Negro History* (January 1963), pp. 1–10, and Arthur S. Link, "Woodrow Wilson: The American as Southerner," *Journal of Southern History* 36, no. 1 (February 1970), pp. 3–17.

7. See Peter Filene, "An Obituary for the 'Progressive Movement,'" *American Quarterly* 22 (Spring 1970), pp. 20–34; Daniel T. Rodgers, "In Search of Progressivism," *Reviews of American History* 10 (December 1982), pp. 114–23.

8. For an overview, see Alonzo L. Hamby, "Progressivism: A Century of Change and Rebirth," in Sidney M. Milkis and Jerome M. Mileur, eds., *Progressivism and the New Democracy* (Amherst: University of Massachusetts Press, 1999), pp. 40–80. For more or less contemporaneous treatments, see, e.g., S. J. Duncan-Clark, *The Progressive Movement: Its Principles and Its Programme* (Boston: Small, Maynard, 1913), chs. 1–2, 17; H. L. Mencken, "Roosevelt: An Autopsy," in *H. L. Mencken: Prejudices, A Selection*, ed. James T. Farrell (New York: Vintage Books, 1958), pp. 47–69; and Merriam, *A History of American Political Theories*, chs. 8–9.

9. James T. Kloppenberg, *Uncertain Victory: Social Democracy and Progressivism in European and American Thought, 1870–1920* (New York: Oxford University Press, 1986), pp. 298–300; Daniel T. Rodgers, *Atlantic Crossings: Social Politics in a Progressive Age* (Cambridge, MA: Belknap Press, 1998), p. 52.

10. Walter Lippmann, *Drift and Mastery* (New York: Mitchell Kennerley, 1914), pp. 152–53; Kloppenberg, *Uncertain Victory*, p. 298.

11. Woodrow Wilson, *A History of the American People*, 5 vols. (New York: Harper & Brothers, 1902), vol. 5, p. 58; Ronald J. Pestritto, *Woodrow Wilson and the Roots of Modern Liberalism* (Lanham, MD: Rowman & Littlefield, 2005), pp. 44–45.

12. Progressive Platform of 1912, in Ronald J. Pestritto and William J. Atto, eds., *American Progressivism: A Reader* (Lanham, MD: Lexington Books, 2008), pp. 273–87, at 274. Wilson, *The New Freedom*, pp. 4, 35, 57. Cf. Morton Keller, *Affairs of State: Public Life in Late Nineteenth Century America* (Cambridge, MA: Belknap Press, 1977), chs. 8, 14.

13. Isaac Kramnick and Theodore J. Lowi, eds., *American Political Thought: A Norton Anthology* (New York: Norton, 2009), pp. 787, 802–3.

14. Theodore Roosevelt, "The Prophecies of Mr. Bryan," in William H. Harbaugh, ed., *The Writings of Theodore Roosevelt* (Indianapolis: Bobbs-Merrill, 1967), p. 48.

15. Quoted in Richard Hofstadter, *The Age of Reform: From Bryan to F.D.R.* (New York: Vintage Books, 1955), p. 132.

16. Kramnick and Lowi, *American Political Thought*, pp. 803–4.

17. As James H. Davis, a Texas Populist, exulted, "The principles of the Declaration of Independence, the Constitution of the United States, and the National Demands of the . . . People's Party . . . contain all the requisite

provisions for the grandest and most perfect civilization." James H. Davis, *A Political Revelation* (1894), in Norman Pollack, ed., *The Populist Mind* (Indianapolis: Bobbs-Merrill, 1967), pp. 203–27, at 204. Cf. Henry Demarest Lloyd, *Wealth Against Commonwealth* (1894): "The Constitution and laws of the United States are, however imperfectly, the translation into the language of politics of doing as you would be done by—the essence of equal rights and government by consent." In Pollack, *The Populist Mind*, pp. 496–534, at 523.

18. See Charles A. Beard, *An Economic Interpretation of the Constitution of the United States* (New York: Free Press, 1986; orig. ed., 1913), and J. Allen Smith, *The Spirit of American Government: A Study of the Constitution: Its Origin, Influence and Relation to Democracy* (New York: Macmillan, 1907).

19. Quoted in Richard Hofstadter, *The American Political Tradition and the Men Who Made It* (New York: Vintage Books, 1974; orig. ed., 1948), p. 316. Hofstadter's debunking of Populism in this book (Bryan lived and died "a provincial politician following a provincial populace in provincial prejudices," p. 265) and in his other famous study, which won the Pulitzer Prize for history in 1956, confirms his own Progressive inclinations, as he admitted. Cf. Hofstadter, *The Age of Reform*, pp. 12–21 and chs. 1–3.

20. Wilson, *The New Freedom*, pp. 37, 42. Thomas Jefferson, *Notes on the State of Virginia*, Query 17, in Merrill D. Peterson, ed., *The Portable Thomas Jefferson*, p. 213. Abraham Lincoln, Speech on the Dred Scott Decision, in *Lincoln: Selected Speeches and Writings*, ed. Don E. Fehrenbacher, (New York, NY: Library of America, 1992), pp. 117–121; Speech at Springfield, Illinois, in Fehrenbacher, ed., *Lincoln: Selected Speeches and Writings*, pp. 277–278. On the views of progress prevailing in America before Progressivism, see the discerning discussion in James Ceaser, *Nature and History in American Political Development: A Debate* (Cambridge: Harvard University Press, 2008).

21. See the thoughtful treatment in Wilson Carey McWilliams, "Standing at Armageddon: Morality and Religion in Progressive Thought," in Sidney M. Milkis and Jerome M. Mileur, eds., *Progressivism and the New Democracy* (Amherst: University of Massachusetts Press, 1999), pp. 103–25. Cf. Alexis de Tocqueville, *Democracy in America*, trans. Harvey C. Mansfield Jr. and Delba Winthrop (Chicago: University of Chicago Press, 2000), pp. 274–88, 407–25, 504–24.

22. *The Federalist Papers*, No. 49, pp. 311–314; but cf. No. 14, pp. 99–100. See Charles R. Kesler, "The Founders and the Classics," in *The American Founding: Essays on the Formation of the Constitution*, ed. J. Jackson Barlow, Leonard W. Levy, and Ken Masugi (New York: Greenwood Press, 1988), pp. 57–90; and Gary Rosen, *American Compact: James Madison and the Problem of Founding* (Lawrence: University Press of Kansas, 1999).

23. Woodrow Wilson, *Congressional Government: A Study in American Politics* (Baltimore: Johns Hopkins University Press, 1981; orig. ed., 1885), pp. 27, 215.

24. Richard T. Ely, *An Introduction to Political Economy* (New York: Chautauqua Press, 1889), pp. 85–86, 315–16; Woodrow Wilson, *The State:*

*Elements of Historical and Practical Politics* (Boston: D. C. Heath, 1895), sec. 23, and cf. 1159: "Society is in no sense artificial; it is as truly natural and organic as the individual man himself. As Aristotle said, 'man is by nature a social animal; his social function is as normal with him as is his individual function.'"

25. See the careful new study by Jason R. Jividen, *Claiming Lincoln: Progressivism, Equality, and the Battle for Lincoln's Legacy in Presidential Rhetoric* (DeKalb: Northern Illinois University Press, 2011), especially chs. 2 and 3; and Merrill D. Peterson, *Lincoln in American Memory* (New York: Oxford University Press, 1994), pp. 141–94.

26. See especially Theodore Roosevelt, "The Heirs of Abraham Lincoln," February 12, 1913, in *The Works of Theodore Roosevelt*, National Edition, ed. Hermann Hagedorn, 20 vols. (New York: Charles Scribner's Sons, 1926), vol. 17, pp. 359–78; and Wilson, "Abraham Lincoln: A Man of the People," in *College and State: Educational, Political, and Literary Papers (1875–1913)*, vol. 2, pp. 83–101, in *The Public Papers of Woodrow Wilson* (New York: Harper & Brothers, 1925), 6 vols., ed. Ray Stannard Baker and William E. Dodd.

27. Wilson, *Congressional Government*, pp. xi, 10; Wilson, *Division and Reunion, 1829–1889* (New York: Longmans, Green, 1906), pp. 216, 287; Theodore Roosevelt, "The New Nationalism," in Harbaugh, ed., *The Writings of Theodore Roosevelt*, pp. 315–33, at 321; Jividen, *Claiming Lincoln*, pp. 34–37, 58–75. America's lack of a proper State had been an element of Hegel's criticism of the U.S. See *The Philosophy of History*, pp. 84–6; and James Ceaser, *Reconstructing America: The Symbol of America in Modern Thought* (New Haven: Yale University Press, 1997), ch. 7.

28. Christopher Jencks and David Riesman, *The Academic Revolution* (Garden City, NY: Doubleday, 1968), pp. 1–8. Jefferson's University of Virginia was exceptional in trying to keep clergymen at arm's length. But the point of his university, as Jefferson saw it, was to educate Virginia's future statesmen, rendering them proper guardians of their fellow citizens' rights—not to turn the graduates into experts whose knowledge gave them a new title to rule their fellow citizens.

29. David Herbert Donald, *Lincoln* (New York: Simon & Schuster, 1995), pp. 29, 460–66.

30. Jencks and Riesman, *The Academic Revolution*, pp. 12–14.

31. Among others who studied in Germany were John Burgess, Simon Patten, Charles Merriam, W. E. B. DuBois, and Walter Weyl. Of Wilson's peers, those who studied exclusively in America included John Dewey and Frederick Jackson Turner. See Jurgen Herbst, *The German Historical School in American Scholarship* (Ithaca, NY: Cornell University Press, 1965), ch. 1; Eldon Eisenach, *The Lost Promise of Progressivism* (Lawrence: University Press of Kansas, 1994); John G. Gunnell, *The Descent of Political Theory* (Chicago: University of Chicago Press, 1993), pp. 24–32.

32. Gunnell, *The Descent of Political Theory*, p. 23; Frederic C. Howe, *Wisconsin: An Experiment in Democracy* (New York: Charles Scribner's Sons, 1912), pp. vii–xii.

33. Howe, *Wisconsin: An Experiment in Democracy*, pp. 38–39.
34. A well-known case of such pressure was Ely's "trial" in 1894 before the Regents of the University of Wisconsin, of all places, on the charge of teaching socialism. See Kloppenberg, *Uncertain Victory*, pp. 210–11, 265–66.
35. Wilson, "The Study of Administration," in Ronald J. Pestritto, ed., *Woodrow Wilson: The Essential Political Writings* (Lanham: Rowman & Littlefield, 2006), pp. 231–248, at 232.
36. Georg Wilhelm Friedrich Hegel, *The Philosophy of History*, trans. J. Sibree (New York: Dover, 1956), pp. 17–19, 29–31.
37. Ibid., pp. 17–20, 31, 33–37.
38. Cf. Wilson, *The New Freedom*, pp. 38–40 with 41–2.
39. Frances Fukuyama, "The End of History?" in *National Interest*, Summer 1989; *The End of History and the Last Man* (New York: Harper Perennial, 1993).
40. Leo Strauss, "Restatement on Xenophon's *Hiero*," in Leo Strauss, *On Tyranny*, rev. and expanded ed., ed. Victor Gourevitch and Michael S. Roth (New York: Free Press, 1991), pp. 177–212, at 212; Harry V. Jaffa, "The End of History Means the End of Freedom," http://www.claremont.org/publications/pub_print.asp?pubid=8.
41. Quoted in Strauss, "Restatement," p. 210.
42. Wilson, *The New Freedom*, p. 33. Cf. *A Crossroads of Freedom: The 1912 Campaign Speeches of Woodrow Wilson*, ed. John Wells Davidson (New Haven, CT: Yale University Press, 1956), p. 245.
43. John Dewey, "The Influence of Darwin on Philosophy," in Kramnick and Lowi, eds., *American Political Thought*, pp. 1030–35, at 1031, 1035.
44. Paul F. Boller Jr., *American Thought in Transition: The Impact of Evolutionary Naturalism, 1865–1900* (Chicago: Rand McNally, 1969), pp. 54–56.
45. Gertrude Himmelfarb, "Varieties of Social Darwinism," in her *Victorian Minds* (New York: Knopf, 1968), pp. 314–32, at 315–16; Obama, Commencement Address at Knox College, Galesburg, Illinois, June 4, 2005, http://obamaspeeches.com/019-Knox-College-Commencement-Obama-Speech.htm; "A Hope to Fulfill": Remarks to the National Press Club, April 26, 2005, http://www.votesmart.org/public-statement/92449/a-hope-to-fulfill.
46. The pejorative use of "Social Darwinism" dates at least to 1906, though the sentiment is present as early as the 1880s. In the chapter titles of his renowned study, Hofstadter calls Sumner a Social Darwinist and Lester Frank Ward, Sumner's leftist counterpart, simply a "Critic." Richard Hofstadter, *Social Darwinism in American Thought* (Boston: Beacon Press, 1955; orig. ed., 1944), pp. 51, 67, 82.
47. For creative euphemisms, see Eric Goldman, *Rendezvous with Destiny* (New York: Knopf, 1953), and John Dewey, *Liberalism and Social Action* (New York: Perigee Books, 1980; orig. ed., 1935).
48. Himmelfarb's essay "Varieties of Social Darwinism" and her *Darwin and the Darwinian Revolution* (Chicago: Ivan R. Dee, 1996; orig. ed., 1959), part 6, remain the best accounts.
49. Quoted in Boller, *American Thought in Transition*, pp. 49–51.
50. Ibid., pp. 64–69.

51. He liked to compare himself to George Washington. As his biographer and erstwhile friend, William Bayard Hale, recalled, "On another occasion, commenting on the curious part the number *thirteen* had played in his own life, he mentioned to his biographer the circumstance that the name, WOODROW WILSON, like that of GEO. WASHINGTON, contained thirteen letters." A peculiar observation, comparing their signatures rather than full names, and neglecting to mention that Wilson didn't call himself Woodrow, dropping his first name Thomas, until he had gotten his B.A. from Princeton and begun his legal studies at the University of Virginia. His Princeton friends knew him as "Tommy." Hale, *The Story of a Style* (New York: Huebsch, 1920), p. 76; John Milton Cooper Jr., *Woodrow Wilson: A Biography* (New York: Knopf, 2009), p. 36.

52. Wilson, *The New Freedom*, pp. 44–48; *Constitutional Government in the United States*, pp. 54–57.

53. Wilson, *Constitutional Government in the United States*, pp. 54–56; *The New Freedom*, pp. 46–47. Wilson consistently disregards the framers' intention to form a national government that would be not merely checked and balanced but also "energetic," republican, and capable of pursuing long-term projects for the public good. See Charles R. Kesler, "Woodrow Wilson and the Statesmanship of Progress," in Thomas B. Silver and Peter W. Schramm, eds., *Natural Right and Political Right: Essays in Honor of Harry V. Jaffa* (Durham, NC: Carolina Academic Press, 1984), pp. 103–27; Kesler, "Separation of Powers and the Administrative State," in Gordon Jones and John Marini, eds., *The Imperial Congress: Crisis in the Separation of Powers* (New York: Pharos Books, 1989), pp. 1–21; Ronald J. Pestritto, *Woodrow Wilson and the Roots of Modern Liberalism*. And see Harvey C. Mansfield Jr., *Taming the Prince: The Ambivalence of Modern Executive Power* (Baltimore: Johns Hopkins University Press, 1993); and William Kristol, "The Problem of the Separation of Powers: *Federalist* 47–51," in Charles R. Kesler, ed., *Saving the Revolution: The Federalist Papers and the American Founding* (New York: Free Press, 1987).

54. Wilson, *Constitutional Government in the United States*, pp. 41–4, 57; Niels Aage Thorsen, *The Political Thought of Woodrow Wilson, 1875–1910* (Princeton: Princeton University Press, 1988), p. 145.

55. Wilson, *The New Freedom*, pp. 28–31, 38–44. Cf. Wilson, *The State*, secs. 1160, 1286.

56. Wilson, *The New Freedom*, pp. 29–30, 44; *The Papers of Woodrow Wilson*, ed. Arthur S. Link, 69 vols. (Princeton, NJ: Princeton University Press, 1966–94), vol. 10, pp. 22–23. On the two Wilsons, see the fine discussion and thorough citations in Pestritto, *Woodrow Wilson and the Roots of Modern Liberalism*, especially pp. 19–23, 253–54. Cf. Hofstadter's conclusion: "The early Wilson made room in his philosophy for change, for reform, as an organic principle, and his ultimate conversion is no more drastic than a change of emphasis." Hofstadter, *The American Political Tradition*, p. 317.

57. Wilson, "A Calendar of Great Americans," in Pestritto, ed., *Woodrow Wilson: The Essential Political Writings*, pp. 81–90 at 85. Cf. Pestritto,

*Woodrow Wilson and the Roots of Modern Liberalism*, pp. 10–13, 51–54, 254–57.

58. Wilson, *Constitutional Government in the United States*, pp. 56–57. Emphasis added. Cf. Hegel, *Elements of the Philosophy of Right*, trans., H. B. Nisbet, ed. Allen W. Wood (Cambridge: Cambridge University Press, 1991), sec. 271, p. 304.

59. On the judicial and jurisprudential revolution, see Bradley C. S. Watson, *Living Constitution, Dying Faith: Progressivism and the New Science of Jurisprudence* (Wilmington, DE: Intercollegiate Studies Institute, 2009); Christopher Wolfe, *The Rise of Modern Judicial Review: From Constitutional Interpretation to Judge-Made Law* (Lanham, MD: Rowman & Littlefield, 1994; rev. ed.); and Gary Jeffrey Jacobsohn, *Pragmatism, Statesmanship, and the Supreme Court* (Ithaca, NY: Cornell, 1977).

60. Wilson, *The State*, sec. 1161; Fukuyama, *America at the Crossroads: Democracy, Power, and the Neoconservative Legacy* (New Haven, CT: Yale University Press, 2007), p. xi; Hegel, *The Philosophy of History*, pp. 96, 98, 419–422. The final quote is from Johann Caspar Bluntschli, who at Heidelberg taught the teachers of Wilson and TR (H. B. Adams and John Burgess, respectively), quoted in Herbst, *The German Historical School in American Scholarship*, p. 120.

61. Edward A. Freeman, "Some Impressions of the United States," *Fortnightly Review* 32 (September 1882), p. 327, quoted in Herbst, *The German Historical School in American Scholarship*, p. 122.

62. See Lawrence H. Fuchs, *American Kaleidoscope: Race, Ethnicity, and the Civic Culture* (Middletown, CT: Wesleyan University Press, 1990), chs. 1-3, and Rogers Smith, *Civic Ideals: Conflicting Visions of Citizenship in U.S. History* (New Haven, CT: Yale University Press, 1999). But cf. Charles R. Kesler, "The Promise of American Citizenship," in Noah Pickus, ed., *Immigration and Citizenship in the Twenty-First Century* (Lanham, MD: Rowman & Littlefield, 1998); Thomas G. West, *Vindicating the Founders: Race, Sex, Class, and Justice in the Origins of America* (Lanham, MD: Rowman & Littlefield, 2000); and Edward J. Erler, John Marini, and Thomas G. West, *The Founders on Citizenship and Immigration: Principles and Challenges in America* (Lanham, MD: Rowman & Littlefield, 2007). The Founders, particularly Jefferson, had their own penchant for Anglo-Saxonism, but the differences are instructive. Jefferson extolled the Saxons' love of liberty and democracy, in contrast to the Normans' addiction to aristocracy and feudalism. In his version of the Robin Hood story, after 1066 the Normans diverted English political development onto an absolutist track, which lasted until the overthrow of the Stuarts and perhaps until the American Revolution. He elevated Saxonism to the level of politics, of republican versus monarchical regimes. Wilson preferred to reduce regime questions to the level of culture or race. The one stressed the role of human freedom in politics, the other the role of fate or historical determinism.

63. Wilson, *The State*, sec. 4. He was drawing on a well-established tradition that included not only Filmer but Fustel de Coulanges and Lewis Henry

Morgan, not to mention John C. Calhoun and the southern defenders of chattel slavery.

64. For example, see Obama, President's Weekly Address, November 24, 2011, http://www.presidency.ucsb.edu/ws/index.php?pid=97313&st=&st1=#axz z1lvulJr36; Remarks and a Question-and-Answer Session in Albuquerque, New Mexico, September 28, 2010, at http://www.presidency.ucsb.edu/ws/index.php?pid=88524&st=&st1=#axzz1lvulJr36.

65. Wilson, *The State*, secs. 1173–79, 1236–37, 1244; Richard T. Ely, *An Introduction to Political Economy*, p. 315.

66. Wilson, *The State*, secs. 1244, 1269. Cf. Hegel, *The Philosophy of History*, pp. 37-53. Many of the early libertarian attacks on "statism" reflect some of the same confusion or ambiguity between the German and English senses of the term. See, e.g., Albert Jay Nock, *Our Enemy, the State* (Auburn, AL: Ludwig von Mises Institute, 2009).

67. Wilson, *The State*, secs. 1173, 1175, 1236-37, 1244-45, 1273-74.

68. Wilson, *The State*, secs. 1255, 1269-71, 1273-74, 1283-86.

69. See John Marini, "Theology, Metaphysics, and Positivism: The Origins of the Social Sciences and the Transformation of the American University," in *Challenges to the American Founding: Slavery, Historicism, and Progressivism in the Nineteenth Century* (Lanham: Lexington Books, 2005), ch. 6; Robert Eden, *Political Leadership and Nihilism* (Tampa: University Presses of Florida, 1983), ch. 1; Tiffany Jones Miller, "John Dewey and the Philosophical Refounding of America," *National Review* (December 31, 2009). For "leadership" in *The Federalist*, see nos. 6, 10, 14, 16, 18, 43, 49, 59, 62, 65, 70, and 85. See also *The Federalist Papers*, Nos. 6, 10, 14, 16, 18, 43, 49, 59, 62, 65, 70, and 85.

70. Wilson, "Leaders of Men," pp. 212–213, 221–222, 224.

71. *Ibid.*, pp. 214, 224.

72. Cf. Wilson, *The State*, sec. 1160.

73. Wilson, "Leaders of Men," pp. 214, 226; and Wilson, "Abraham Lincoln: A Man of the People," in Link, ed., *Papers*, vol. 19, pp. 30–59.

74. Wilson, "Leaders of Men," pp. 215, 223, 225, 226.

75. John Wells Davidson, ed., *A Crossroads of Freedom*, p. 187; Wilson, *The New Freedom*, p. 73.

76. See *The Federalist Papers*, Nos. 6, 16, 26, 35, and 46.

77. Jaffa, "The End of History Is the End of Freedom"; Wilson, quoted in Kesler, "Woodrow Wilson and the Statesmanship of Progress," pp. 123 and 127n48.

### Chapter 3: Franklin D. Roosevelt and the Rise of Liberalism

1. *Time*, November 24, 2008; Peter Beinart, "The New Liberal Order," *Time*, November 13, 2008, http://www.time.com/time/magazine/article/0,9171,1858873,00.html.

2. Obama, *The Audacity of Hope*, pp. 176–77; "A Hope to Fulfill": Remarks to the National Press Club, April 26, 2005, http://www.votesmart.org/public-statement/92449/a-hope-to-fulfill.

3. Ibid.

4. Ernest K. Lindley, "Symposium: Early Days of the New Deal," in Morton J. Frisch and Martin Diamond, eds., *The Thirties: A Reconsideration in the Light of the American Political Tradition* (DeKalb: Northern Illinois University Press, 2010; orig. ed., 1968), pp. 131–32. See Lindley, *Franklin D. Roosevelt: A Career in Progressive Democracy* (Indianapolis: Bobbs-Merrill, 1931).

5. Franklin D. Roosevelt, Speech before the 1932 Democratic National Convention, in *The Public Papers and Addresses of Franklin Delano Roosevelt*, vol. 1, p. 648; Commonwealth Club Address, *The Public Papers and Addresses of Franklin Delano Roosevelt*, vol. 1, p. 750. Cf. his Address at the Jackson Day Dinner, Washington, D.C., January 8, 1938: "If the cataclysm of the World War had not stopped [Wilson's] hand, neither you nor I would today be facing such a difficult task of reconstruction and reform." *The Public Papers and Addresses of Franklin D. Roosevelt*, 1938 volume, ed. Samuel I. Rosenman (New York: Macmillan, 1941), p. 41; and see p. 614. FDR echoed Wilson's own account in his Second Inaugural Address of his turn from domestic to foreign policy.

6. Frisch and Diamond, eds., *The Thirties*, p. 129; Hofstadter, *The American Political Tradition*, pp. 311, 379; James MacGregor Burns, *The Lion and the Fox* (New York: Harcourt Brace Jovanovich, 1956), p. 198; Paul K. Conkin, *The New Deal* (New York: Thomas Crowell, 1967), p. 13; and Kesler, "The Public Philosophy of the New Freedom and the New Deal," in Robert Eden, ed., *The New Deal and Its Legacy: Critique and Reappraisal* (New York: Greenwood Press, 1989), pp. 155–66, at p. 160.

7. Rexford G. Tugwell, "The New Deal: The Progressive Tradition," *Western Political Quarterly* (September 1950), pp. 395–96. Cf. Tugwell, "The Progressive Orthodoxy of Franklin D. Roosevelt," *Ethics*, vol. 64, no. 1 (October 1953), pp. 1–23, at 2–3, 14–22.

8. Robert Eden, "On the Origins of the Regime of Pragmatic Liberalism: John Dewey, Adolf A. Berle, and FDR's Commonwealth Club Address of 1932," in *Studies in American Political Development* 7 (Spring 1993), pp. 74–150. Eden seems uncertain at times whether FDR or Berle was in the driver's seat. On necessity and executive power, see Harvey C. Mansfield Jr., *Taming the Prince: The Ambivalence of Modern Executive Power* (New York: Free Press, 1989).

9. Cf. Hofstadter, *The Age of Reform*, pp. 277–86.

10. William Allen White, *The Autobiography of William Allen White* (New York: Macmillan, 1946), pp. 632–33, quoted in Eden, "Origins of Pragmatic Liberalism," p. 89; and Eden, pp. 89–99.

11. *The Public Papers and Addresses of Franklin D. Roosevelt*, 1938 volume, ed. Samuel I. Rosenman (New York: Macmillan, 1941), pp. xxviii–xxxiii. For the six speeches during the primary campaigns, see items 80, 90, 100, 113, 132, and 143. See also item 137.

12. Quoted in Ronald D. Rotunda, *The Politics of Language: Liberalism as Word and Symbol* (Iowa City: University of Iowa Press, 1986), pp. 64, 72; see also ibid., ch. 4. The relevant texts of the debate over liberalism are now conveniently available in Gordon Lloyd, ed., *The Two Faces of Liberalism: How the Hoover-Roosevelt Debate Shapes the 21st Century* (Salem, MA:

M & M Scrivener Press, 2006).

13. Rotunda, *The Politics of Language*, chs. 2, 3; Edward A. Stettner, *Shaping Modern Liberalism: Herbert Croly and Progressive Thought* (Lawrence: University Press of Kansas, 1993), pp. 1–7, 110–13.

14. After TR's defeat in 1912, a new Progressive Party emerged in 1924. Its presidential candidate, Robert M. La Follette, was defeated by Calvin Coolidge.

15. See Rotunda, *The Politics of Language*, pp. 10–11, 90.

16. Wilson, *Constitutional Government in the United States*, p. 57; *The New Freedom*, p. 177.

17. Lloyd, ed., *The Two Faces of Liberalism*, pp. 96–97, 287, 291.

18. Ibid., pp. 292–95.

19. See Bruce Ackerman, *We the People*, vol. 1, *Foundations* (Cambridge, MA: Belknap Press, 1993).

20. Rosenman, ed., *The Public Papers and Addresses of Franklin D. Roosevelt*, 1938 volume, p. 586.

21. Cf. Walter Rauschenbusch, *Christianizing the Social Order* (New York: Macmillan, 1912), pp. 41–42: "The chief purpose of the Christian Church in the past has been the salvation of individuals. But the most pressing task of the present is not individualistic. Our business is to make over an antiquated and immoral economic system; to get rid of laws, customs, maxims, and philosophies inherited from an evil and despotic past; to create just and brotherly relations between great groups and classes of society. . . . The Christian Church in the past has taught us to do our work with our eyes fixed on another world and a life to come. But the business before us is concerned with refashioning the present world, making this earth clean and sweet and habitable."

22. Lloyd, ed., *The Two Faces of Liberalism*, pp. 103, 161, 309.

23. See James Ceaser, *Presidential Selection: Theory and Development* (Princeton, NJ: Princeton University Press, 1979); Sidney Milkis, *The Modern Presidency and the Transformation of the American Party System* (New York: Oxford University Press, 1992).

24. Quoted in Hofstadter, *The Age of Reform*, p. 279; Hofstadter, *The American Political Tradition*, pp. 334–37.

25. Rosenman, ed., *Public Papers and Address of Franklin D. Roosevelt*, 1938 volume, p. 585; Lloyd, ed., *The Two Faces of Liberalism*, p. 310.

26. Consider Cass R. Sunstein, *The Second Bill of Rights: FDR's Unfinished Revolution and Why We Need It More than Ever* (New York: Basic Books, 2004).

27. "Is There a Jefferson on the Horizon?" *New York Evening World*, December 3, 1925, reprinted in Basil Rauch, ed., *The Roosevelt Reader: Selected Speeches, Messages, Press Conferences, and Letters of Franklin D. Roosevelt* (New York: Holt, Rinehart & Winston, 1957), pp. 43–47, at 47; Lloyd, ed., *The Two Faces of Liberalism*, p. 115. Cf. FDR's speech at Columbus, Ohio, on August 20, 1932: "I believe that our industrial and economic system is made for individual men and women, and not individual men and women for the benefit of the system." *Public Papers and Addresses of Franklin D. Roosevelt*, vol. 1 (New York: Random House,

1938), pp. 684–92, at p. 680. By contrast, Wilson was cool toward Jefferson on account of his states' rights views and penchant for abstract theory. Cf. Niels Aage Thorsen, *The Political Thought of Woodrow Wilson, 1875–1910* (Princeton, NJ: Princeton University Press, 1988), pp. 221–22, 237; and Pestritto, *Woodrow Wilson and the Roots of Modern Liberalism,* pp. 104–5, 118, 174–76. Jefferson was, "though a great man, not a great American," according to Wilson. "A Calendar of Great Americans," in Pestritto, ed., *Woodrow Wilson: The Essential Political Writings,* p. 86. But cf. *The New Freedom,* p. 55, where he says the same thing about Hamilton.

28. Cf. Wilson, *Constitutional Government in the United States,* chs. 1–2.

29. Cf. ibid., pp. 5–10. The inaptness of Magna Carta to the American case was emphasized by Hamilton in *Federalist* 84: "It has been several times truly remarked that bills of rights are, in their origin, stipulations between kings and their subjects, abridgements of prerogative in favor of privilege, reservations of rights not surrendered to the prince. Such was MAGNA CHARTA, obtained by the barons, swords in hand, from King John. . . . It is evident, therefore, that, according to their primitive signification, they have no application to constitutions, professedly founded upon the power of the people and executed by their immediate representatives and servants. Here, in strictness, the people surrender nothing, and as they retain everything, they have no need of particular reservations." *The Federalist Papers,* p. 512.

30. See the penetrating discussion in Eden, "On the Origins of the Regime of Pragmatic Liberalism," pp. 106–9, 122–39.

31. The most notable instance is FDR's discussion of the Court-packing plan. His claim was that government is "a three horse team provided by the Constitution to the American people so that their field may be ploughed. The three horses are, of course, the three branches of government," he said in his Fireside Chat on March 9, 1937. "Two of the horses are pulling in unison today; the third is not. . . . It is the American people themselves who expect the third horse to pull in unison with the other two." Rosenman, ed., *The Public Papers and Addresses of Franklin D. Roosevelt,* 1937 volume, pp. 123–24; cf. p. 116. In his view, American government is purely a matter of power, horsepower in fact! The judicial, legislative, and executive powers do not have distinctive qualities that need to be respected and combined in republican government.

32. Eden, "On the Origins of the Regime of Pragmatic Liberalism," pp. 106–9, 122–39.

33. Roosevelt, "New Conditions Impose New Requirements upon Government and Those Who Conduct Government." Campaign Address on Progressive Government at the Commonwealth Club, San Francisco, September 23, 1932, in *The Public Papers and Addresses of Franklin Delano Roosevelt,* vol. 1, pp.742–55; "We Are Fighting to Save a Great and Precious Form of Government for Ourselves and the World," Acceptance of the Renomination for the Presidency, Philadelphia, June 27, 1936, in *Public Papers,* vol. 5, pp. 230–35; Second Inaugural Address, "I See One-Third of a Nation Ill-Housed, Ill-Clad, Ill-Nourished," January 20, 1937, in *Public Papers,* 1937 volume, pp. 1–5.

34. See the excellent discussion in Thomas G. West, *Vindicating the Founders* (Lanham, MD: Rowman & Littlefield, 2000), ch. 2.

35. For "war against poverty," see his Radio Address to the 1940 Democratic National Convention, July 19, 1940, in Rosenman, ed., *The Public Papers and Addresses of Franklin D. Roosevelt*, 1940 volume, pp. 293–302.

36. For further evidence of economic contradictions in New Deal policies, see Burton Folsom Jr., *New Deal or Raw Deal?* (New York: Threshold Editions, 2008), and Jim Powell, *FDR's Folly: How Roosevelt and His New Deal Prolonged the Great Depression* (New York: Crown Forum, 2003).

37. The classic statement is Isaiah Berlin, "Two Concepts of Liberty," in *Liberty: Incorporating Four Essays on Liberty*, ed. Henry Hardy (New York: Oxford University Press, 2002), pp. 166–217.

38. See the excellent discussions in Harry V. Jaffa, *A New Birth of Freedom: Abraham Lincoln and the Coming of the Civil War* (Chicago: University of Chicago Press, 2000), ch. 2; and Harvey C. Mansfield Jr., "Responsibility Versus Self-Expression," in Robert A. Licht, ed., *Old Rights and New* (Washington: AEI Press, 1993), pp. 96–111. On natural rights and the virtues, ancient and modern, see Mark Blitz, *Duty Bound: Responsibility and American Public Life* (Lanham, MD: Rowman & Littlefield, 2005).

39. So far as I know, FDR never claimed that the new rights were natural rights. Something like this claim was raised by a few American political scientists influenced by the Hegelian school. For example, Theodore D. Woolsey, the long-serving president of Yale College, in his *Political Science or the State* (New York: Charles Scribner's Sons, 1886), vol. 1, ch. 1, sec. 11: "We mean then by natural rights those which, by fair deduction from the present physical, moral, social, religious characteristics of man, he must be invested with, and which he ought to have realized for him in a jural society, in order to fulfill the ends to which his nature calls him." Woolsey contrasts this definition with the "heathenish" sense of natural rights arising from the "uncontrolled liberties" of man in the state of nature. Socioeconomic rights, in Roosevelt's sense, were discussed by some Progressives but with nothing like the prominence and the Jeffersonian perfume he gave them. Consider, e.g., Walter Weyl, *The New Democracy* (New York: Macmillan, 1914), p. 161: "The inner soul of our new democracy is not the unalienable rights, negatively and individualistically interpreted, but those same rights . . . extended and given a social interpretation." When so interpreted, they become "public trusts."

40. Roosevelt, "New Conditions Impose New Requirements Upon Government and Those Who Conduct Government," ibid.

41. Quoted in Folsom, *New Deal or Raw Deal?*, p. 35. See in general Folsom, ch. 3; Peter Temin, *Lessons from the Great Depression* (Cambridge, MA: MIT Press, 1989); Gene Smiley, *Rethinking the Great Depression* (Chicago: Ivan Dee, 2002); and Richard Vedder, "Explaining the Great Depression," *Claremont Review of Books* 10, no. 2 (Spring 2010), pp. 44–48. The economic data on underconsumption are analyzed with devastating effect in Thomas B. Silver, *Coolidge and the Historians* (Durham, NC: Carolina Academic Press, 1982), pp. 124–36.

42. Folsom, *New Deal or Raw Deal?*, ch. 10; and on the inherent barriers to redistribution, see Voegeli, *Never Enough*, ch. 3.

43. See Howard Zinn, ed., *New Deal Thought* (Indianapolis: Bobbs-Merrill,

1966), p. xvii.

44. Cf. Voegeli, *Never Enough*, pp. 97–102.

45. Rodgers, *Atlantic Crossings*, pp. 414–15; quotation at 415.

46. Roosevelt would never count on such motives alone, of course. Hence his famous remark on requiring every worker to pay Social Security tax on every paycheck: Those taxes "are political all the way through," he said, because "with those taxes in there, no damn politician can ever scrap my social security program." Quoted in Folsom, *New Deal or Raw Deal?*, p. 117.

47. Irving Kristol, "Ten Years in a Tunnel," in Frisch and Diamond, eds., *The Thirties*, pp. 20–24. James MacGregor Burns looked forward to a Children's Bill of Rights as well as an Artists' Bill of Rights, for example. See James MacGregor Burns and Stewart Burns, *A People's Charter: The Pursuit of Rights in America* (New York: Vintage, 1993).

48. Rosenman, ed., *The Public Papers and Addresses of Franklin D. Roosevelt*, 1937 volume, p. 135.

### Chapter 4: Lyndon B. Johnson and the Politics of Meaning

1. Obama, *The Audacity of Hope*, pp. 36-37; Steven F. Hayward, *The Age of Reagan: The Fall of the Old Liberal Order, 1964–1980* (Roseville, CA: Prima, 2001), p. xxx.

2. Obama, *The Audacity of Hope*, pp. 29–32. His "Sixties" extended well beyond the decade, which was not unusual.

3. Ibid., pp. 31, 34–35.

4. Ibid., pp. 27, 38, 155, 224–25, 253.

5. Hayward, *The Age of Reagan*, p. xxiv.

6. Richard N. Goodwin, *Remembering America: A Voice from the Sixties* (Boston: Little, Brown, 1988), pp. 258, 269–70.

7. William E. Leuchtenburg, "Lyndon Johnson in the Shadow of FDR," in *The Great Society and the High Tide of Liberalism*, ed. Sidney M. Milkis and Jerome M. Mileur (Amherst: University of Massachusetts Press, 2005), pp. 185-213, at 185, 204-05; Goodwin, *Remembering America*, p. 259.

8. Lyndon B. Johnson, Special Message to the Congress Proposing a Nationwide War on the Sources of Poverty, March 16, 1964, http://www.presidency.ucsb.edu/ws/index.php?pid=26109#axzz1oOQAXaMV; Remarks at the University of Michigan, May 22, 1964, http://www.presidency.ucsb.edu/ws/index.php?pid=26262#axzz1oOQAXaMV. Milkis, "Lyndon Johnson, The Great Society, and the Modern Presidency," in Milkis and Mileur, *ibid.*, pp. 1–49, at 7, 9–11, 40.

9. Johnson, Remarks at the University of Michigan, May 22, 1964.

10. Daniel Patrick Moynihan, *Coping: Essays on the Practice of Government* (New York: Random House, 1973), p. 8; Moynihan, *Maximum Feasible Misunderstanding: Community Action and the War on Poverty* (New York: Free Press, 1969), p. 99; emphasis in the original, as quoted in Hayward, *The Age of Reagan*, pp. 11, 17. Samuel Beer quoted in Milkis, "Lyndon Johnson . . . ," p. 29.

11. Johnson, Remarks at the University of Michigan, May 22, 1964; Goodwin,

*Remembering America*, pp. 278–81.

12. Johnson, Remarks at the University of Michigan, May 22, 1964; Goodwin, *Remembering America*, pp. 425–26; Milkis, "Lyndon Johnson . . . ," p.33.

13. Johnson, Remarks at the University of Michigan, May 22, 1964; Goodwin, *Remembering America*, pp. 425–26; Milkis, "Lyndon Johnson," p. 33.

14. Goodwin, *Remembering America*, p. 426; Johnson, Special Message to the Congress Proposing a Nationwide War on the Sources of Poverty, March 16, 1964. Cf. Gareth Davies, *From Opportunity to Entitlement: The Transformation and Decline of Great Society Liberalism* (Lawrence: University Press of Kansas, 1996), p. 242.

15. Richard T. Ely, *Social Aspects of Christianity, and Other Essays*, new and enlarged ed. (New York: Thomas Y. Crowell, 1889), pp. 123–24.

16. Simon N. Patten, *The New Basis of Civilization*, p. 208.

17. William Leuchtenburg, *Franklin D. Roosevelt and the New Deal* (New York: Harper Perennial, 1963), p. 345; Rodgers, *Atlantic Crossings*, pp. 446–84.

18. Goodwin, *Remembering America*, pp. 272–78.

19. See Robert A. Caro, *Master of the Senate*, (New York: Knopf, 2002) pp. 136–38, 587, 589–90.

20. See Tulis, *The Rhetorical Presidency*, pp. 161–72.

21. Hayward, *The Age of Reagan*, pp. 82–85.

22. Johnson, Commencement Address at Howard University: To Fulfill These Rights, June 4, 1965.

23. Hayward, *The Age of Reagan*, p. 86.

24. Harvey C. Mansfield Jr., "The Legacy of the Late Sixties," in Stephen Macedo, ed., *Reassessing the Sixties: Debating the Political and Cultural Legacy* (New York: Norton, 1997), pp. 21–45, at 42–43.

25. Hayward, *The Age of Reagan*, pp. 87–95.

26. See John Adams Wettergreen, "Bureaucratizing the American Government," in Gordon S. Jones and John A. Marini, eds., *The Imperial Congress: Crisis in the Separation of Powers* (New York: Pharos Books, 1988), ch. 3.

27. Hayward, *The Age of Reagan*, p. 7. See Theodore J. Lowi, *The End of Liberalism: The Second Republic of the United States* (New York: Norton, 2009); Morris Fiorina, *Congress: Keystone of the Washington Establishment* (New Haven, CT: Yale University Press, 1989); John Marini, *The Politics of Budget Control: Congress, The Presidency, and the Growth of the Administrative State* (New York: Taylor & Francis, 1992).

28. Hayward, *The Age of Reagan*, p. 67.

29. The Port Huron Statement, in Kramnick and Lowi, *ibid.*, pp. 1290-1301, at 1290-96. Cf. Allan Bloom, *The Closing of the American Mind: How Higher Education Has Failed Democracy and Impoverished the Souls of Today's Students* (New York: Simon & Schuster, 1987), pp. 194-216.

30. Cf. Harry V. Jaffa, *A New Birth of Freedom*, pp. 95-96.

31. See Mary Parker Follett, *The New State* (New York: Longman, Green, 1918), p. 138.

32. Hugh Heclo, "Sixties Civics," in Milkis and Mileur, *ibid.*, pp. 53-82, at 54-55, 64-65, 69, 71.

33. Jimmy Carter, Speech on Energy Policy and National Goals, July 15, 1979,

www.presidency.ucsb.edu/ws/index.php?pid=32596&st=&st1=#axzzlpVsf889c.

## Chapter 5: Obama and the Crisis of Liberalism

1.  Athena Jones, "Obama: Change Comes from Me," MSNBC, November 26, 2008; Remarks in Columbia, Missouri, October 30, 2008, http://www.presidency.ucsb.edu/ws/index.php?pid=84665&st=&st1= #axzz1lvulJr36; Bret Stephens, "Is Obama Smart?" *Wall Street Journal,* August 9, 2011, http://online.wsj.com/article/SB100014240531 11904140604576495932704234052.html; Sarah Parnass, "Obama Biography: 'I'm LeBron, Baby,'" ABC News, June 13, 2011, at http://abcnews.go.com/blogs/politics/2011/06/obama-biography-im-lebron-baby/; Interview with Steve Kroft, *60 Minutes,* CBS, December 9, 2011, http://www.cbsnews.com/8301-18560_162-57341024/ interview-with-president-obama-the-full-transcript/.

2.  Tulis, *The Rhetorical Presidency,* pp. 45–67. See also James W. Ceaser, Glen Thurow, Jeffrey Tulis, and Joseph M. Bessette, "The Rise of the Rhetorical Presidency," *Presidential Studies Quarterly* 11, no. 2 (Spring 1981), pp. 158–171.

3.  Tulis, *The Rhetorical Presidency,* pp. 27–59, 95–116, 189–202.

4.  Wilson, "Cabinet Government," in Link, ed., *The Papers of Woodrow Wilson,* volume 5 (Princeton, NJ: Princeton University Press, 1968), pp. 389–90, and vol. 2 (Princeton, NJ: Princeton University Press, 1967), p. 500, and "Leaders of Men," quoted in Tulis, *The Rhetorical Presidency,* pp. 126, 129–30.

5.  Franklin Delano Roosevelt, Second Inaugural Address, "I See-One-Third of a Nation Ill-Housed, Ill-Clad, Ill-Nourished," January 20, 1937, in Rosenman, ed., *Public Papers and Addresses,* 1937 volume, pp. 1–5; Bill Clinton, Second Inaugural Address, January 20, 1997, http://www.bartleby. com/124/pres65.html; Obama, Commencement Address at Knox College, Galesburg, Illinois, June 4, 2005, at http://obamaspeeches.com/019-Knox-College-Commencement-Obama-Speech.htm.

6.  The scientific study of divine and human political visions reached its peak in the Middle Ages. See the discussion of prophecy in Maimonides, *Guide of the Perplexed,* vols. 1–2, trans. Schlomo Pines (Chicago: University of Chicago Press, 1974).

7.  The verses come from O'Shaughnessy's "Ode," in his collection *Music and Moonlight* (1874). The whole poem, a celebration of poets and dreamers as the legislators of mankind, is worth reading.

8.  Franklin Delano Roosevelt, "Inaugural Address," March 4, 1933, in Rosenman, ed., *Public Papers and Addresses,* vol. 1, pp. 11–16.

9.  Article II, section 1, of the Constitution begins: "The executive power shall be vested in a President of the United States of America."

10. See Clinton L. Rossiter, *Constitutional Dictatorship: Crisis Government in the Modern Democracies* (Princeton, NJ: Princeton University Press, 1948), chs. 16–18.

11. See Obama, Remarks Following the Iowa Caucuses, January 3, 2008,

http://www.presidency.ucsb.edu/ws/index.php?pid=76232&st=&st1=#ax
zz1lvulJr36; Remarks Following the New Hampshire Primary, January 8,
2008, http://www.presidency.ucsb.edu/ws/index.php?pid=62272&st=&st1
=#axzz1lvulJr36.

12. Obama, *Dreams from My Father*, p. xvii. Much light has been shed on the
fictional strains in the book by David Remnick's biographical researches in
*The Bridge* and by Stanley Kurtz's in his *Radical-in-Chief: Barack Obama
and the Untold Story of American Socialism* (New York: Threshold Editions, 2010).

13. Ezra Klein, "Obama's Gift," January 4, 2008, http://prospect.org/article/
obamas-gift.

14. Obama, "Address Before a Joint Session of Congress on Job Growth,"
September 8, 2011, http://www.presidency.ucsb.edu/ws/index.php?pid=966
61&st=&st1=#axzz1lvulJr36.

15. As William Kristol pointed out, the military "weeds out those not up
to the task" and imposes "tough and demanding training," not to mention "fitness and discipline and good character." Military life is therefore
quite unlike modern liberalism. "But welfare state liberalism is all about
scratching each other's backs; nanny state liberalism is all about rubbing
each other's backs; and entitlement state liberalism is all about stroking
each other's backs. None is about protecting each other's backs—let alone
driving away our enemies and turning around bravely to face the future."
William Kristol, "Obama: Follow the Example of the Military?," January
25, 2012, http://www.weeklystandard.com/blogs/obama-follow-military-
example_618484.html.

16. Harvey C. Mansfield, Jr., "What Obama Isn't Saying," *Weekly Standard*,
February 8, 2010.

17. In *The Audacity of Hope*, Obama writes that in a democracy "no law is
ever final, no battle truly finished." James Kloppenberg comments, "which
is why philosophical pragmatism and deliberative democracy go hand in
hand. Principled partisans of pragmatism and democracy are committed to
debate, experimentation, and the critical reassessment of results"—except
when they're not. Or perhaps Obama is not as pragmatic as he and Kloppenberg think. James T. Kloppenberg, *Reading Obama* (Princeton, NJ:
Princeton University Press, 2011), pp. 169–70.

18. Obama is presumably dating the goal of national health care to its mention
in the Progressive Party platform of 1912.

19. Obama, Remarks to the American Medical Association National Conference in Chicago, Illinois, June 15, 2009, http://www.presidency.ucsb.edu/
ws/index.php?pid=86285&st=&st1=#axzz1mEsej8iI; Remarks at a Virtual
Town Hall and a Question-and-Answer Session on Health Care Reform
in Annandale, Virginia, July 1, 2009, http://www.presidency.ucsb.edu/ws/
index.php?pid=86368&st=&st1=#axzz1mEsej8iI.

20. See, e.g., Sally C. Pipes, *The Pipes Plan: The Top Ten Ways to Dis-
mantle and Replace Obamacare* (Washington: Regnery Books, 2012)
and *The Truth About Obamacare* (Washington: Regnery Books, 2010);

Grace-Marie Turner, James C. Capretta, Thomas P. Miller, and Robert E. Moffit, *Why Obamacare is Wrong for America* (New York: Broadside Books, 2011); and preceding Obamacare but still relevant, David Gratzer, *The Cure: How Capitalism Can Save American Health Care* (New York: Encounter Books, 2008), and Regina Herzlinger, *Who Killed Health Care?* (New York: McGraw-Hill, 2007).

21. Obama, Remarks on House of Representatives Passage of Health Care Reform Legislation, March 21, 2010, http://www.presidency.ucsb.edu/ws/?pid=87654#axzz1lvulJr36.

22. Nancy Pelosi, Remarks at the 2010 Legislative Conference for the National Association of Counties, March 9, 2010, http://www.democraticleader.gov/news/speeches?id=0249; David Cho et al., "Lawmakers Guide Dodd-Frank bill for Wall Street Reform into Homestretch," June 26, 2010, http://www.washingtonpost.com/wp-dyn/content/article/2010/06/25/AR2010062500675_pf.html; Chris Good, "Baucus Defends Health Care, Didn't Read the Entire Bill," August 25, 2010, http://www.theatlantic.com/politics/archive/2010/08/baucus-defends-health-care-didnt-read-the-entire-bill/62030/.

23. John Adams Wettergreen, "Bureaucratizing the American Government," ibid.

24. As noted above, Obama was originally opposed to the individual mandate. A friendly observer hailed his pragmatic statesmanship: "Offering an ambitious health care plan, he would not require adults to purchase health insurance. His goal is to make health care available, not to force people to buy it—a judgment that reflects Obama's commitment to freedom of choice, his pragmatic nature (an enforcement question: Would those without health care be fined or jailed?), and his desire to produce a plan that might actually obtain a consensus." Cass R. Sunstein, "The Empiricist Strikes Back," *New Republic*, September 10, 2008. Hadley Arkes argues trenchantly that the deepest objections to the president's health care law arise from its violations of natural rights, in "Natural Rights Trump Obamacare, or Should," *First Things*, December 2011, pp. 41–46.

25. Cf. George F. Will, "Government by the 'Experts,'" *Washington Post*, June 10, 2011, http://www.washingtonpost.com/opinions/government-by-the-experts/2011/06/09/AGpU1KPH_story.html.

26. See Marvin Olasky, *The Tragedy of American Compassion* (Wheaton, IL: Crossway Books, 1992).

27. Obama, *The Audacity of Hope*, p. 53.

28. Lincoln, Letter to H. L. Pierce and Others, April 6, 1859, in *The Collected Works of Abraham Lincoln*, ed. Roy P. Basler (New Brunswick, NJ: Rutgers University Press, 1953), vol. 3, p. 376; Jefferson, Letter to Henry Lee, May 8, 1825, and Letter to Roger Weightman, June 24, 1826, in Merrill D. Peterson, ed., *Thomas Jefferson: Writings* (New York: Library of America, 1984), pp. 1501, 1517. For a commentary, see Harry V. Jaffa, *A New Birth of Freedom* (Lanham, MD: Rowman & Littlefield, 2000), ch. 2.

29. Cf. Steven Hayward, "The Liberal Republicanism of Gordon Wood," *Claremont Review of Books* 7, no. 1 (Winter 2006), pp. 27–30; and Kloppenberg, *Reading Obama*, pp. 41–44, 52–54. See Gordon Wood, *The Creation*

*of the American Republic, 1776–1787* (Chapel Hill: University of North Carolina Press, 1969), and for his second thoughts, *The Radicalism of the American Revolution* (New York: Vintage, 1993).

30. Obama, "A Hope to Fulfill," Remarks of Senator Barack Obama at the National Press Club, April 26, 2005, http://obamaspeeches.com/014-National-Press-Club-Obama-Speech.htm; Remarks Following the Wisconsin Primary, February 19, 2008, http://www.presidency.ucsb.edu/ws/index.php?pid=76558&st=&st1=#axzz1lvulJr36; Remarks in St. Paul, Minnesota, Claiming the Democratic Presidential Nomination Following the Montana and South Dakota Primaries, June 3, 2008, http://www.presidency.ucsb.edu/ws/index.php?pid=77409&st=&st1=#axzz1lvulJr36; Address Before a Joint Session of Congress on the State of the Union, January 24, 2012, http://www.presidency.ucsb.edu/ws/index.php?pid=99000#axzz1lvulJr36; Kloppenberg, *Reading Obama*, pp. 89–110, 139–40.

31. Obama, Comments at Presidential Debate at Belmont University in Nashville, Tennessee, October 7, 2008, http://www.presidency.ucsb.edu/ws/index.php?pid=84482&st=&st1=#axzz1lvulJr36.

32. Obama, *The Audacity of Hope*, pp. 55, 92.

33. Obama, *The Audacity of Hope*, p. 95.

34. Obama, "A More Perfect Union": Address at the National Constitution Center in Philadelphia, March 18, 2008, http://www.presidency.ucsb.edu/ws/index.php?pid=76710&st=&st1=#axzz1lvulJr36; cf. Kloppenberg, *Reading Obama*, p. 209, and cf. 212: "When Wright froze American racism into a fixed feature of the national culture, he was betraying two principles Obama embraced: democracy and historicism."

35. Kloppenberg, *Reading Obama*, pp. 209–12.

36. Ibid., pp. 78–80.

37. Martin Luther King, *Why We Can't Wait* (New York: Signet Classic, 2000; orig. ed., 1963), pp. 70–72, 76–77. Obama, *The Audacity of Hope*, p. 93.

38. Obama, "A More Perfect Union": Address at the National Constitution Center in Philadelphia, March 18, 2008; Obama, *The Audacity of Hope*, pp. 93–96. Obama echoes, and radicalizes, Woodrow Wilson's distinction between the Founders as time-bound theorists and as competent statesmen.

39. Obama, *The Audacity of Hope*, pp. 97–98.

40. Ibid., p. 98.

41. Richard Rorty, *Contingency, Irony, and Solidarity* (Cambridge, Eng.: Cambridge University Press, 1989).

42. See, for example, Walter Russell Mead, "Beyond the Blue Part One: The Crisis of the American Dream," *American Interest*, January 29, 2012, http://blogs.the-american-interest.com/wrm/2012/01/29/beyond-blue-part-one-the-crisis-of-the-american-dream/.

43. Yuval Levin, "Beyond the Welfare State," *National Affairs*, Spring 2011, pp. 21–38, 30, 32.

44. Clive Crook, "A Tainted Victory," *Atlantic*, March 22, 2010, http://www.theatlantic.com/politics/archive/2010/03/a-tainted-victory/37811. Quoted in William Voegeli, *Never Enough*, foreword to the paperback edition (New York: Encounter Books, 2012).

# Index

Wilson, Woodrow *(cont.)*
Jefferson and, 250n 62, 253n 27;
leadership and, 90–100, 119, 161,
175, 180; liberalism of, xxiv, 169;
libertarianism of, 74; "living constitution" and, 75–76, 84, 88, 89,
90; mentors and teachers of, 47;
modernity and, 44; New Freedom
and, xxiii, 43, 88, 105–6, 112, 123–
24; new political ideas of, 32–33,
69–70, 74, 81, 82–84, 86–87, 105,
128, 132, 134, 249n 56; patriarchalism and, 80, 250n 63; politics of
vision and, 186; political liberalism
of, 34; Populism and, 43, 99–100;
presidency and, 32, 33–34, 121,
181; on presidency and foreign
relations, 244n 3; as Princeton's
president, 69, 153; Progressivism
and, 31, 33, 69–88, 120, 158–59,
230; property rights and, 140;
quoting from *Through the Looking
Glass*, 63; on race and government,
78–79; relativism and, 227; Social
Darwinism and, 66; socialism and,
87–88; southern prejudices and,
35, 37–38, 73; speechmaking (rhetorical presidency) of, 179, 182;
Spencer's ideas and, 68; the State
and, 81, 82–86, 120; transformation of constitutional system and,
45–47, 49, 69–88, 232, 249n 53;
Washington and, 249n 51; World
War I and, 188, 252n 5
Winfrey, Oprah, 14
Witherspoon, John, 50–51
Wood, Gordon, 214
Woodrow, Thomas, 34
Woolsey, Theodore D., 255n 39
World War I, 79, 102, 106, 188, 252n 5;
Treaty of Versailles, 108, 124, 244n 3
Wright, Jeremiah, xviii, 219, 261n 34;
America as racist, 219, 220, 221;
"audacity of hope" and, 18